Understanding The Bu
Regulations - 2nd Editi

Understanding The Building Regulations - 2nd Edition

Simon Polley

London and New York

First published 1995 by E & FN Spon
Reprinted 1996, 1997, 2000

Second edition published 2001 by Spon Press
11 New Fetter Lane, London EC4P 4EE

Simultaneously published in the USA and Canada
by Spon Press
29 West 35th Street, New York, NY 10001

Reprinted 2002

Spon Press is an Imprint of the Taylor & Francis Group

© 2001 Simon Polley

Typeset in 10/12 pt Palatino by Deerpark Publishing Services Ltd, Shannon
Printed and bound in the United Kingdom at the University Press, Cambridge

British Library Cataloguing in Publication Data
A catalogue record for this book is available from the British Library

Library of Congress Cataloguing in Publication Data
A catalogue record for this book has been requested

ISBN 0-419-24720-3

Contents

Preface

The purpose of this book is to introduce the reader to the current system of building control in England and Wales, based on the Building Regulations 1991 and all the supporting Approved Documents. The Building Regulations (Amendment) Regulations 1992–1999 and Approved Documents 1992–2000 are included in the text.

The book starts with a brief history lesson in building control and how it has developed over the years. Chapters are then devoted to the Building Regulations and each of the Approved Documents. The application of Building Regulation Requirements and the guidance contained in Approved Documents is discussed and illustrated in a straightforward and logical manner to enable the text to be utilized as a reference source for members of the design and construction team and those who require a knowledge of building control. For information and reference, the actual wording of each Building Regulation Requirement has been stated at beginning of each chapter.

Since the book represents a simplification of the Building Regulations and Approved Documents it should not be regarded as a replacement but as a standalone support text to the original documents. For further detailed advice and guidance the services of an approved inspector, local authority building control department or a building regulation consultancy should be sought.

Simon Polley MBEng MRICS MIFS MIFireE
Managing Director, BRCS (Building Control) Limited
7 Rivermead North, Bishop Hall Lane, Chelmsford
Essex CM1 1PD

Acknowledgements

The author would like to thank the following people for their assistance and support in producing this book:

Frank Robinson BEng (Hons), MBEng MRICS, Director, BRCS (Building Control) Limited.

Cartoons and illustrations by 'Bill' Brignell, Cartoonist and Illustrator, 25 Town End Field, Witham, Essex CM8 1EU.

CAD drawings and diagrams by Paul Scott MRICS ABEng, P.A. Scott Associates, Chartered Building Surveyors, The Gate House, 116 Rainsford Road, Chelmsford, CM1 2QL.

Introduction

In London in 1189, Henry Fitz Ailwyn was appointed mayor. His London Assize was virtually the first London Building Act – the first Building Regulations. As an example, if neighbours agreed to build a party wall between their adjoining properties it had to be 3 ft thick and 16 ft high. Unfortunately, no powers existed to enforce this requirement or impose penalties.

Archaeological ruins and debris show that, prior to this date, failures occurred, and that building codes were used mainly as deterrents. In 2000 BC, for example, in Egypt, the ultimate deterrent was introduced as a regulation: if a man died because of a building failure then the builder himself would be held liable and put to death.

In Roman times, rules and regulations were drawn up and enforcement was introduced to try and reduce the possibility of failure, which has been the main theme of regulations and enforcement to date.

Returning to this country, the seventeenth century saw the first Acts to cover the whole of England, although it is the year 1666 that will be remembered, especially by Charles II. In the early hours of Sunday 1 September the Great Fire of London started at the King's baker in Pudding Lane: four-fifths of the City was destroyed.

In the aftermath, the City appointed four surveyors to draft Regulations, which were duly embodied in the Rebuilding Act of 1667. They included:

- the utilization of four different 'purpose groups';
- minimum storey heights;
- party/external wall thicknesses;
- space separation;
- rainwater pipe provisions.

These regulations for the first time were well detailed, and City Viewers – the first building control surveyors – were appointed to enforce compliance. The late nineteenth century saw the emergence, as a consequence of the Public Health Act 1875, of the two distinct systems of building control, with the District Surveyor responsible for the London County Council area, and local authorities (under the Public Health Act) for the remainder of the country. The model by-laws used were rather basic, reflecting the comparatively simple construction techniques, although importantly they did require that suitable plans be submitted for local authority approval.

In 1881 the average number of persons per dwelling was eight, and despite the many regulations, domestic building standards both in London and across the country were poor. Not until after the First World War did things start to improve. The complexity of building had started to increase with the wide use of structural steelwork, in lieu of cast iron, and the introduction of reinforced concrete. This brought about a more mathematical analysis of building design, a fact that all later legislation took into account.

The principal Building Acts of the time were the London Building Act 1930, amended in 1935 and 1939, and for the remainder of England and Wales the Public Health Act 1936, from which a single series of model by-laws was issued.

After the Second World War was over, the availability of a large range of codes of practice and British Standard specifications mirrored the ever-changing growth of building technology. The Building By-laws of 1953 first gave the option of 'deemed to satisfy' requirements, and on 1 February 1966 the first national Building Regulations came into operation. These were made under the Public Health Act 1961 and included provisions for:

- structural fire precautions;
- requirements for division or compartment walls;
- fire protection to structural elements;
- sound insulation to walls and floors of dwellings;
- minimum stairway dimensions;
- 'Zones of open space';
- a table of exempted buildings.

The Building Regulations 1972 (basically a metric reissue) and the Building Regulations 1976 were to follow. In April 1980 fees for the submission of plans were introduced for the first time under the Building (Prescribed Fees) Regulations.

In February 1981 a Command Paper – *The future of building control in England and Wales* – was published. It contained the Government's proposals for major changes to building control, and saw the light of day in the form of the Building Act 1984. This consolidated nearly all the previous legislation covering building control and included new proposals for optional privatization, streamlining of the system and redrafting of the Building Regulations. The private option for building control first appeared under the Building (Approved Inspector) Regulations 1985.

The Building Regulations 1985 reflected the contents of the 1976 regulations, but their form and arrangement was drastically altered. The Schedule 1 Requirements were written in general terms requiring reasonable standards of health and safety for persons in or about the building. Reference was made to supporting Approved Documents which, with the exception of B1 – Means of escape, were not legally enforceable, as various ways could exist to show compliance.

Following a review of the 1985 regulations, by the Department of the Environment, the Building Regulations 1991 came into force on 1 June 1992, with more than half of the Approved Documents being revised. The Building Regulations (Amendment) Regulations 1994 brought about changes to specific regulations and requirements, which included new approved documents for Parts F – Ventilation and L – Conservation of fuel and power, operable from 1 July 1995.

Since 1995 the now Department of the Environment, Transport and Regions has issued a number of amendment regulations. The Building Regulations (Amendment) (No 2) Regulations 1999 resulted in the 2000 edition of Approved Document B. Approved Documents K and N and the approved document to support Regulation 7 have been revised, and the 1999 edition of Approved Document M now applies to dwellings for the first time.

Two non-departmental approved documents have been published, *Timber intermediate floors for dwellings* (TRADA 1992) and *Basements for dwellings* (British Cement Association 1997). With the introduction of the Building (Local Authority Charges) Regulations 1998 local authority building control can now determine their own fee scales for building regulation submissions. To act as a guide for the submission of building regulation applications the *Manual to the Building Regulations* has also been re-introduced.

In January 1997 the private option for building control was expanded by the arrival of Approved Inspectors with the ability to operate in the commercial sector. The Building (Approved Inspectors, etc.) Regulations 2000 consolidated previous regulations and incorporated minor amendments and additional requirements. The building regulations themselves were also consolidated and the Building Regulations 2000 came into force on the 1 January 2001.

CHAPTER 1

The Building Regulations 2000

The Building Regulations 2000 are a Statutory Instrument, 2000/2531, made under specific sections of the Building Act 1984. They impose requirements on people carrying out certain building work within England and Wales. It is important to note that compliance must be shown with these regulations and not necessarily the contents of the Approved Documents, which are purely to give 'practical guidance with respect to the requirements of any provision of building regulations' [Section 6 of the Building Act 1984], and their use is not therefore mandatory.

The actual Regulations, which are discussed below, are split into five parts concluding with three schedules:

- Schedule 1 – the technical requirements expressed in functional terms to which building work must comply;
- Schedule 2 – exempt buildings and work as referred to in Regulation 9;
- Schedule 3 – the list of regulations that have been revoked as referred to in Regulation 24.

Regulation 1: Citation and commencement

Regulation 2: Interpretation

A number of terms used within the regulations are explained. The significant ones are reiterated for information:

Building – any permanent or temporary building (or part of a building) but not any other kind of structure or erection.
Controlled service or fitting – a service or fitting where Part G, H or J imposes a requirement.
Dwelling – includes a dwelling-house and a flat.
Dwelling-house – does not include a flat or a building containing a flat.

European technical approval – a favourable technical assessment of the fitness for use of a construction product for the purposes of the Construction Products Directive, issued by a European Technical Approval issuing body.

Flat – separate and self-contained premises, including a maisonette, constructed or adapted for residential use and forming part of a building divided horizontally from some other part.

Floor area – the aggregate area of every floor in a building or extension, calculated by reference to the finished internal faces of the walls enclosing the area, or if at any point there is no such wall, by reference to the outermost edge of the floor.

Height – the height of the building measured from the mean adjoining outside ground level to a level half the vertical height of the roof or to the top of any walls or parapets, whichever is the higher.

Public building – consisting of or containing:

(a) a theatre, public library, hall or other place of public resort,
(b) a school or other educational establishment [not exempted under section 4(1)(a) of the Building Act 1984], or
(c) a place of public worship,

but excluding a shop, storehouse or warehouse, or a dwelling to which members of the public are occasionally admitted.

Regulation 3: Meaning of building work

One of the first tasks with respect to any proposal is to establish whether it is **building work** and hence requires a submission under the Building Regulations. Building work is defined as:

(a) the erection or extension of a building;
(b) the provision or extension of a controlled service or fitting in or in connection with a building;
(c) the material alteration of a building, or a controlled service or fitting;
(d) work required for a material change of use;
(e) insertion of cavity wall insulation material; or
(f) work involving the underpinning of a building.

A **material alteration** occurs if the work, or any part of it, would at any stage result in a non-compliance, where it previously complied or, if it did not comply with a relevant requirement, by it becoming **more unsatisfactory**. Only Requirements A, B1, B3, B4, B5 and M, relating to structure, fire safety and disabled access, are relevant to a material alteration. Examples would include an opening in a load-bearing wall, erection of new internal partitions giving rise to increased means of escape travel distances, or the removal of a disabled toilet or access ramp.

Regulation 4: Requirements relating to building work

Building work, as established above, must be carried out in accordance with the relevant requirements listed in Schedule 1. These include the structural stability of the building, means of escape and fire safety, resistance to moisture, ventilation arrangements, drainage, stair design, thermal insulation and facilities for disabled people. To comply with one requirement should not cause a non-compliance with another requirement.

Where compliance with the relevant requirements of Schedule 1 was not originally shown, the building work shall not make the situation **more unsatisfactory** than it was before the work was carried out.

Regulation 5: Meaning of material change of use

From a building regulation point of view a **material change of use** only occurs where the use of a building is changed to:

(a) a dwelling;
(b) contain a flat;
(c) a hotel or boarding house;
(d) an institution;
(e) a public building (as defined in Regulation 2);
(f) a building no longer exempt under Classes I-VI of Schedule 2;
(g) a building containing a greater or lesser number of dwellings.

Regulation 6: Requirements relating to material change of use

Depending on the actual change of use that will take place, certain **relevant requirements** of Schedule 1 will need to be applied:

- B1, B2, B3, B4(2), B5, F1-F2, G1-G2, H4, J1-J3 and L1 for all cases;
- A1-A3 for cases (c), (d), (e) and (f) described in Regulation 5 above (e.g. a barn conversion to a hotel);
- B4(1) in the case of a building over 15 m in height;
- C4 for case (a) (i.e. a dwelling); and
- E1-E3 for cases (a), (b) or (g) (e.g. a flat conversion).

Regulation 7: Materials and workmanship

The functionally written regulation states that building work complying with the relevant requirements of Schedule 1 shall be carried out

with adequate and **proper materials** and in a **workmanlike manner.**
Regulation 7 is supported by its own approved document, the contents
of which are discussed later in the book.

Regulation 8: Limitation on requirements

This important regulation clarifies that compliance with Parts A-K and
N is limited to secure **reasonable** standards of health and safety for
persons in or about buildings (and others who may be affected by
buildings or matters connected with them). Therefore only a reasonable duty of care rests with the designer and/or builder as far as
building regulations are concerned. This also applies from the view-
point of the local authority, who need only establish compliance with
the requirements as limited above. This was the subject of case law,
Murphy v *Brentwood District Council* [1988], where losses of a financial
nature were not recognized.

Regulation 9: Exempt buildings and work

Schedule 2 lists the classes of buildings and work which are exempt
from the application of the building regulation requirements:

Class I   Buildings controlled under other legislation (i.e. the Explosives Acts 1875 and 1923, Nuclear Installations Act 1965 and
the Ancient Monuments and Archeological Areas Act 1979).

Class II   A detached building not frequented by people where
isolated by at least 1.5 times the building height from a
controlled building or boundary.

Class III   Greenhouse. Agricultural building, including a building
principally for the keeping of animals (e.g. a stable),
where it is not a dwelling, is isolated by at least 1.5 times
the building height from sleeping accommodation, and is
provided with a fire exit within 30 m. A greenhouse or
agricultural building would not be exempt if it was principally used for retailing, packing or exhibiting.

Class IV   Temporary buildings (e.g. mobile homes), not intended to
remain erected for more than 28 days.

Class V   Ancillary buildings (i.e. site buildings containing no sleeping accommodation).

Class VI   Small detached buildings containing no sleeping accommodation are exempt up to 15 m^2 in floor area or up to 30 m^2
where the building is at least 1 m from a boundary or if
constructed of substantially non-combustible materials. A
nuclear fallout shelter is also exempt subject to a maximum
30 m^2 floor area and it being isolated from another building
or structure by the depth of the excavation plus 1 m.

Class VII Extensions up to 30 m^2 (i.e. conservatory, porch, covered yard, covered way or carport with at least two open sides), although conservatories and porches incorporating glazing should satisfy the requirements of Part N.

Regulation 10: The Metropolitan Police Authority

The Metropolitan Police Authority gain exemption from the procedural requirements as a public body.

Regulation 11: Power to dispense with or relax requirements

All the requirements contained within Schedule 1 are written in a functional form and should not therefore be relaxed. The **Dispensation** of a particular requirement may be reasonable in the circumstances whereupon application can be made to the local authority. If refused, a right of appeal exists, under Section 39 of the Building Act 1984, to the Department of the Environment.

A **Determination** can also be requested from the Department of the Environment, under Section 16(10) of the Building Act 1984, so as to resolve a question between the controlling authority and the applicant. Under this procedure the work or element of work in question should not have commenced at the time of making the application for a determination. A fee for the application is applicable, which stands at half the plan fee (excluding the VAT) with minimum and maximum limits of £50 and £500 respectively, all as stated in the Building (Local Authority Charges) Regulations 1998.

Regulation 12: Giving of a building notice or deposit of plans

A person intending to carry out building work or to make a material change of use must make a submission in one of two forms where the local authority system of building control is to be utilized (please also refer to Figure 1.1 which also illustrates the alternative route using an **Approved Inspector**):

(a) give to the local authority a **building notice** (Regulation 12), or
(b) deposit **full plans** (Regulation 13); this will be necessary for a building put to a **designated use** under the Fire Precautions Act 1971 (i.e. hotel, boarding house, factory, office, shop and railway premises) or a **relevant use** as a work place subject to the Fire Precautions (work place) Regulations 1997 (as amended).

A submission will not be necessary for the installation of a heat-producing gas appliance if the installer is approved under the Gas Safety (Installation and Use) Regulations 1998.

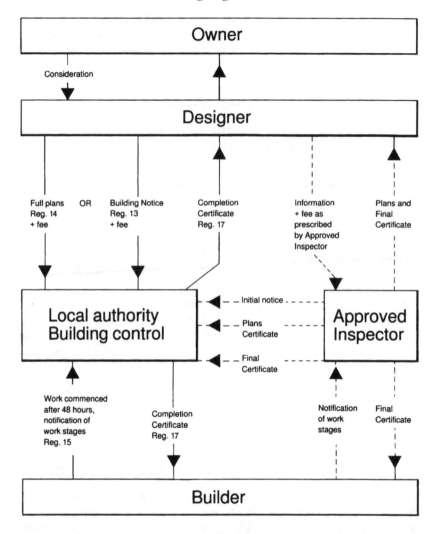

Figure 1.1 Building regulation submission flowchart. *Note:* Dashed line route represents the alternative method of gaining design and work approval utilizing an **Approved Inspector**. For specific guidance, reference should be made to the Building (Approved Inspector) Regulations 2000 (as amended) or an Approved Inspector should be consulted (e.g. BRCS (Building Control) Limited).

Regulation 13: Particulars and plans where a building notice is given

The giving of a building notice represents a simple method of notifying the local authority that building work or a change of use is proposed. It requires the submission of the following information:

- the name and address of the person intending to carry out the work and signed by him or on his behalf;
- a statement that the notice is given in accordance with Regulation 12(2)(a);
- a description of the proposal;
- the location and proposed use of the building; and

For an erection or extension of a building the following are also needed:

- a 1:1250 or greater scale plan showing the size, position and curtilage boundaries of the building and other buildings and streets within that curtilage;
- the number of storeys (including basement storeys) in the building;
- drainage provisions and precautions to be taken if building is over a public sewer (reference Sections 21 and 18 of the Building Act 1984 respectively); and
- steps to be taken to ensure compliance with any local enactment.

For the insertion of cavity wall insulation the statement needs to confirm:

- the name and type of insulation material proposed;
- whether the material conforms to a national standard of a member state, name of any issuing body and the requirements approved; and
- whether the installer is approved, and the name of the body which issued the approval.

Where the provision of an unvented hot water storage system is proposed the statement needs to confirm:

- the name, make, model and type of system;
- whether the system is approved or certified to satisfy Requirement G3; and
- whether the installer has a current registered operative identity card.

To enable the local authority to verify compliance with the regulations they may also request in writing the submission of further plans within a specified time period. These could include: additional structural calculations; detailed floor layout drawings for means of escape purposes; or the specification of a disabled person passenger lift.

It should be noted that a building notice, or the plans submitted with it, are not treated as a formal submission for building regulation approval under Section 16 of the Building Act 1984. The choice could therefore be made to submit a full plans application, which would give the benefit of an Approval Notice and a Completion

Certificate once the plans and work on site had been found to comply with the regulations. Note that unless the work has commenced a building notice becomes of no effect after 3 years.

Regulation 14: Full plans

As previously stated, the deposit of a full plans application offers the dual benefits of an Approval Notice and Completion Certificate, where so requested under Regulation 17. A certain degree of information is required to enable the authority to check full compliance with the Building Regulations:

- the name and address of the person intending to carry out the work and signed by him or on his behalf;
- a statement that the notice is given in accordance with Regulation 12(2)(b);
- a description of the proposal;
- the location and proposed use of the building;
- a 1:1250 or greater scale plan showing the size, position and curtilage boundaries of the building and other buildings and streets within that curtilage;
- the number of storeys (including basement storeys) in the building;
- drainage provisions and precautions to be taken if building is over a public sewer (reference Sections 21 and 18 of the Building Act 1984 respectively);
- steps to be taken to ensure compliance with any local enactment;
- a statement to confirm if the building will be put to a relevant use.

For the insertion of cavity wall insulation the statement needs to confirm:

- the name and type of insulation material proposed;
- whether the material conforms to a national standard of a member state, name of any issuing body and the requirements approved; and
- whether the installer is approved, and the name of the body which issued the approval.

Where the provision of an unvented hot water storage system is proposed the statement needs to confirm:

- the name, make, model and type of system;
- whether the system is approved or certified to satisfy Requirement G3; and
- whether the installer has a current registered operative identity card.

In addition to all of the above, '**any other plans which are necessary to show that the work would comply with these regulations**' shall also be deposited. In this regard it is very important to identify the extent of the information that the local authority will need to verify compliance with the Regulations. For example, reams and reams of builder's specification notes or working details can be very time-consuming to produce and are unlikely to be required. However, it will be necessary to provide detailed floor layouts, plans and sections to an appropriate scale and specific details where compliance with a particular regulation needs to be shown, e.g. damp-proof course arrangements at an external wall/roof abutment or the provision of sound insulation around an internal soil and vent pipe.

The submission of full plans shall be in duplicate with two additional copies provided where Part B – Fire safety applies, but with the exception of houses and flats. This allows the local authority to consult the local fire authority, and is explained later in the book.

Once full plans been deposited and acknowledged by the local authority they must approve or reject them within 5 weeks, or 2 months if an extension of time is agreed. Any rejection notice must state where the plans do not conform to the Regulations or where additional information is necessary.

A further benefit of the full plans procedure is that plans may be passed conditionally by one of these two methods:

- where the plans show a contravention the local authority may approve them subject to the necessary correction being made, e.g. a damp-proof course not indicated or a particular door not specified as a fire door; and/or
- the local authority may approve the plans subject to further specific plans being deposited at a later date, normally before that specific element of work has commenced, e.g. design and details of timber trussed rafters or the design and specification of an active fire safety feature. This method can enable the plans to be dealt with in stages.

In both cases the local authority is not obliged to use them and the applicant must give written agreement.

Regulation 15: Notice of commencement and completion of certain stages of work

This regulation states where notice needs to be given to the local authority by the person carrying out building work and before that work has commenced:

- 2 days' notice for commencement (i.e. any period of 48 hours commencing at midnight excluding weekends and public/bank holidays). Note that the full plans submission or building notice needs to have been with the authority for this time before work can commence.

- 1 day's notice for excavations, foundations, DPCs, oversites and drainage (i.e. any period of 24 hours commencing at midnight excluding weekends and public/bank holidays).
- 5 days' notice for completion, occupation and completion of drainage work.

The manner in which notice is given is not stated, but could include writing, facsimile, telephone or other verbal communication. Where notice has not been given and the work has been covered up, the local authority can require the work to be cut into, laid open or pulled down where necessary to establish compliance. For example, a trial hole and core sample may be deemed necessary to ascertain the depth and quality of foundation concrete already cast and not seen by the local authority building control officer.

It is important to note that building work may be commenced, constructed and completed without the benefit of a Building Regulation approval or having to construct the building work in accordance with the approved plan. This is subject to giving the notices described above and constructing the building work itself so as to comply with the relevant requirements of Schedule 1 to the Building Regulations 1991 (as amended). This regulation does not apply for work supervised by an Approved Inspector.

Regulation 16: Energy rating

Where a new dwelling is created either by new building work or a material change of use it shall be given an **energy rating**. This rating is to be calculated under a procedure approved by the Secretary of State and be notified to the local authority on or before the dwelling has been occupied and/or completed. A notice should also be affixed in the dwelling stating this energy rating.

Regulation 17: Completion certificates

The local authority has an obligation to issue a completion certificate when notified of completion and where:

- the building is put to a relevant use (certificate need only cover Part B); or
- requested with the Full Plans submission.

The local authority will need to take **reasonable steps** to ensure that the building works show compliance with the relevant requirements of Schedule 1. A certificate issued under this regulation can be regarded as evidence, but not conclusively, that the relevant requirements specified in the certificate have been complied with.

Regulation 18: Testing of drains and private sewers

The local authority itself may undertake tests of drains or private sewers so as to establish compliance with Part H – Drainage and waste disposal. For work to or entry into a public sewer the local drainage authority must be contacted.

Regulation 19: Sampling of material

As with drainage, the local authority itself may take samples for testing so as to establish compliance with the regulations.

Regulation 20: Supervision of building work otherwise than by local authorities

If the building work is to be supervised by an **Approved Inspector** (or public body), where an initial notice and final certificate are given, then Regulations 12, 15, 16, 17, 18 and 19 shall not apply. Where the services of an Approved Inspector are to be utilized then reference should be made to the Building (Approved Inspectors etc.) Regulations 2000, please also refer to Figure 1.1.

Regulation 21: Unauthorized building work

To regularize a situation where unauthorized building work has taken place, on or after 11 November 1985 and a submission has not been deposited, the owner or applicant may request in writing that the local authority issue a **Regularization Certificate**. The submission of the following information will be necessary:

- a statement that the application is made in accordance with Regulation 21,
- a description of the unauthorized work,
- a plan of the unauthorized work, if reasonably practicable,
- a plan of any additional work, if reasonably practicable, where necessary to enable compliance to be shown with the relevant requirements applicable at the time of construction.

To enable the local authority to establish that compliance has been shown they may require the applicant to take **reasonable steps** in laying open the unauthorized work. This could include, for example, the excavation of trial holes to verify foundation depths, the testing of drainage, or the taking of building material samples for analysis. It may also be necessary to relax or dispense with a requirement, as described under Regulation 11 above.

Once the local authority have taken all reasonable steps, notified the applicant accordingly and satisfied themselves that the unauthorized building work now complies with the building regulations they *may* issue a Regularization Certificate. A certificate issued under this regulation can be regarded as evidence, but not conclusively, that the relevant requirements specified in the certificate have been complied with.

It should be noted that this Regulation is without prejudice to any action the local authority may take under Section 36 of the Building Act 1984 relating to the **removal or alteration of offending work**.

Regulation 22: Contravention of certain regulations not to be an offence

Namely that an offence under Section 35 of the Building Act 1984 is not applicable for Regulation 17 – Completion certificates.

Regulation 23: Transitional provisions

Regulation 24: Revocations

Schedule 3 lists the regulations that are revoked, including the Building Regulations 1991.

CHAPTER 2

Approved Document to support Regulation 7: Materials and workmanship

> Regulation 7 states:
>
> Building work shall be carried out:
> (a) with adequate and proper materials which:
>   (i) are appropriate for the circumstances in which they are used;
>   (ii) are adequately mixed or prepared; and
>   (iii) which are applied, used or fixed so as to adequately perform the functions for which they are designed; and
> (b) in a workmanlike manner.

The functionally written regulation is basically saying that building work, to comply with the relevant requirements of Schedule 1, shall be carried out using appropriate workmanship, with materials that are:

- of a suitable nature and quality in relation to the purposes and conditions of their use; and
- adequately mixed or prepared; and
- applied, used or fixed so as to perform adequately the functions for which they are intended.

The 1999 edition of this Approved Document acknowledges our relationship with the rest of Europe and the obligations concerning the Construction Products Directive and use of the EC mark. The approved document is split into two sections reflecting the factors listed in the regulation.

Note that the reference to **materials** includes those naturally occurring, e.g. stone, timber and thatch, and products, components, fittings, items of equipment and excavation backfill in connection with building work. The appropriate use of recycled and recyclable materials should now be considered. It is also important to bear in mind the contents of Regulation 8, which clarifies the limited standards required to comply with the Building Regulations and the fact that no continuing control of materials in use is enforced after the work has been completed.

SECTION 1: MATERIALS

To establish the fitness of a material one of a number of methods can be chosen:

- **CE marks** – These denote compliance with a harmonized European Standard or Technical Approval. The controlling authority must accept a material bearing this mark as being fit for its purpose, subject to it being in a satisfactory condition and used appropriately. The onus of proof is therefore with the local building control authority (or Approved Inspector), who should notify the trading standards officer in any particular case.
- **British Standards** – An appropriate British Standard may be referred to or an **equivalent** European Standard.
- **Other national and international technical specifications** – Conformity to the national technical specifications of other Member States.
- **Technical approvals** – Justification may be available from a national or European Certificate issued by a European Technical Approvals issuing body.
- **Independent certification schemes** – Many schemes exist within the UK. Under the Construction Products Directive, certification by an approved body in another Member State shall be accepted by the controlling authority. As with the CE mark the onus of proof is therefore with the local building control authority (or Approved Inspector).
- **Tests and calculations** – The UKAS Accreditation Scheme for Testing Laboratories is an example, as would be an equivalent Member State scheme. This would ensure that the tests, calculations or other means are carried out or undertaken in accordance with agreed criteria.
- **Past experience** – An existing building, where the material in question has been used, may verify its suitability in use and over a reasonable time span.
- **Sampling** – Reference should be made to Regulation 17, which clarifies local authority powers to take and sample materials.

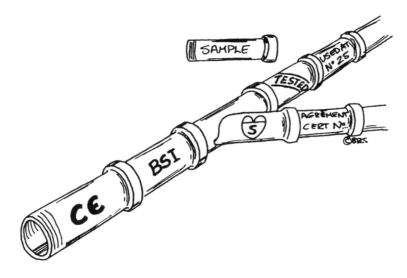

Short-lived materials may be regarded as unsuitable, owing to their potential deterioration over a short period of time if not correctly maintained. If accessibility is possible for inspection, maintenance and repair, then a particular material used in a particular location may be acceptable. This would be subject to the consequences of failure not being serious to persons in or about the building, e.g. an external cladding system up to a modest height where cladding panels could be removed for inspection periodically. It may also be acceptable to use a short-lived material that is not accessible but where the risk to health and safety is low, e.g. a single-layer roofing membrane to a storage building.

The **moisture resistance** of any material, in relation to condensation, rain and snow or from the ground, needs careful consideration. The material should therefore be treated or protected from the effects of moisture or for the building construction to resist the passage of moisture to the material.

Materials in the ground or foundations need to resist attack from substances in the ground or subsoil, e.g. sulphates (cross-reference should be made to Section 2 of Approved Document C).

Materials susceptible to changes in their properties
High-alumina cement is an example which may be used as a heat-resisting material but not for structural work, including foundations. Other examples include certain stainless steel, structural silicone sealants and intumescent paints.

In certain geographical areas of England softwood timber used for roof construction and within the roof void should be treated to prevent infestation by **house longhorn beetle**. The risk areas are as follows:

- the boroughs of Bracknell Forest, Elmbridge, Guildford (excluding the former borough of Guildford), Spelthorne, Surrey Heath, Rushmoor (district of Farnborough only), Waverley (excluding parishes of Godalming and Haslemere) and Woking;
- the districts of Hart (excluding the former urban district of Fleet) and Runnymede;
- the Royal Borough of Windsor and Maidenhead (parishes of Old Windsor, Sunningdale and Sunninghill only).

Note that the above guidance will ultimately move to Approved Document A.

SECTION 2: WORKMANSHIP

To establish the adequacy of a particular method of workmanship a number of methods are available, which are outlined as follows.

- **Standards** – An appropriate British Standard Code of Practice may be referred to, or an equivalent national technical specification of other Member States (e.g. BS 8000: *Workmanship on building sites*).
- **Technical approvals** – Justification may be available from a national or European Certificate issued by a European Technical Approvals issuing body. In this case it will be for the person carrying out the work to show that the method of workmanship will offer an equivalent level of protection and performance.
- **Management systems** – Justification of quality may be possible by the utilization of a scheme showing compliance with BS EN / ISO 9000
- **Past experience** – An existing building, where the workmanship method in question has been used, may verify its suitability to perform the desired function.
- **Tests** – Reference should be made to Regulation 16, which clarifies local authority powers to test drainage.

The approved document concludes with an appendix containing abbreviations, a glossary and a list of the 18 States within the European Economic Area:

Austria	Italy
Belgium	Luxembourg
Denmark	Netherlands
Finland	Norway (not in the EU)
France	Portugal
Germany	Spain
Greece	Sweden
Iceland (not in the EU)	Switzerland (not in the EU)
Ireland	United Kingdom.
(EU – European Union)	

Approved Document A: Structure

Approved Document A – Structure is the first in a series of documents approved by the Secretary of State to offer **practical guidance** on compliance with the Building Regulations 2000. Approved Document A specifically offers guidance on the functional requirements A1, A2 and A3 contained in Part A of Schedule 1 to the Regulations.

In general the guidance contained within the approved document is relatively straightforward, with reliance placed on the vast range of British Standards available to the designer. The opening Sections 1–4 can be used to verify compliance with Requirements A1 and A2.

REQUIREMENT A1: LOADING

The building shall be constructed so that the combined dead, imposed and wind loads are sustained and transmitted by it to the ground:

(a) safely; and
(b) without causing such deflection or deformation of any part of the building, or such movement of the ground, as will impair the stability of any part of another building.

In assessing whether a building complies with the above regard shall be had to the imposed and wind loads to which it is likely to be subjected in the ordinary course of its use for the purpose for which it is intended.

REQUIREMENT A2: GROUND MOVEMENT

The building shall be constructed so that ground movement caused by:

(a) swelling, shrinkage or freezing of the subsoil; or
(b) land-slip or subsidence (other than subsidence arising from shrinkage), in so far as the risk can be reasonably foreseen,

will not impair the stability of any part of the building.

To take account of potential ground movement, buildings should be constructed to transmit loads safely to the ground and not impair the stability of other buildings. The structural safety of a building therefore depends on the successful interrelationship between design and construction, taking into account these particular aspects:

- loading from dead, imposed and wind loads (taking into account possible dynamic, concentrated and peak load effects);
- properties of materials;
- design analysis;
- details of construction;
- workmanship; and
- safety factors (taking into account all the above aspects).

With these aspects in mind the guidance put forward in Sections 1-4 is now discussed.

Section 1: Sizes of structural elements for certain residential buildings and other small buildings of traditional construction

This section opens with a list of definitions, which apply throughout the section:

Buttressing wall – a wall providing full-height lateral support to another wall perpendicular to it.

Dead load – the weight of all walls, permanent partitions, floors, roofs and finishes including services, and all other permanent construction.

Imposed loads – the weight of movable partitions, distributed, concentrated, impact, inertia and snow loads (i.e. due to the proposed occupancy or use) but excluding wind loads.

Pier – a thickened section of wall at intervals to provide lateral support.

Spacing – centre-to-centre longitudinal spacing of adjacent timber members.

Span – centre-to-centre distance between two adjacent supports or bearings (or between faces of bearings where applicable).

Supported wall – a wall provided with lateral support from a buttress wall, pier, chimney, floor or roof members.

Wind load – the load due to the effect of wind pressure or suction.

Section 1A: Basic requirements for stability

- These are to be read in conjunction with Sections 1B and 1C.
- Trussed rafters and traditional roofs, which are not resistant to instability, should be braced in accordance with BS 5268: Part 3: 1985.
- If a roof is sufficiently braced and anchored, and walls are designed in accordance with Section 1C, there is no need to take account of wind pressure or suction.

Table 3.1 Clear spans for certain timber members in single-family houses

Member	Dead load (kN/m²)	Span (m) for member size:		
		50 × 97 / 50 × 100	50 × 147 / 50 × 150	50 × 220 / 50 × 200
Rafter (at 16°)	<1.0	2.18	3.25	–
Purlin	<1.00	–	–	2.25
Ceiling joist	<0.50	1.89	3.19	5.14
Binder	<0.50	–	1.92	2.64
Sheeting purlin	<0.75	1.24	1.85	2.46
Floor joist	<1.25	1.74	2.81	4.07
Flat roof joist	<1.00	1.81	3.04	4.65

Notes: Roof imposed load at 1.00 kN/m²; for imposed snow loads refer to Diagram 2 of the Approved Document or BS 6399: Part 3.
All members at 400 mm centres, 1500 mm centres for purlins and binders. Grade of timber SC3.

Section 1B: Sizes of certain timber floor, ceiling and roof members in single-family houses

- This section applies to single-family houses not more than three storeys high.
- Limits for dead and imposed loads are given and common species/ grades which may be used are listed.
- Appendix A of the approved document contains a series of clear span tables for timber members; some specimen values are given in Table 3.1 for information.
- Double joists may be needed where a floor supports a bath.
- Suitable solid or herringbone strutting should be provided at spans over 2.5 m and 2 rows of strutting over 4.5 m.
- Notches should be no deeper than 0.125 x joist depth, not cut closer to the support than 0.07 x span or further than 0.25 x span.
- Holes should be no greater than 0.25 x joist depth, drilled at neutral axis, spaced at least 3 x hole diameter and located between 0.25 and 0.4 x span from the support.
- Rafters should only have notches for birdsmouthed bearings.

Section 1C: Thickness of walls in certain small buildings

- This section applies to residential buildings up to three storeys high and small single-storey non-residential buildings and annexes; size limitations and conditions for use of the guidance are stated in the Approved Document.
- The thickness of coursed brickwork or blockwork solid walls (paragraph 1C6 of the Approved Document refers) should be at least 1/16 of the storey height and accord with guidelines in Table 3.2.
- Uncoursed stone, flints etc. walls should be at least 1.33 times the thickness determined by paragraph 1C6 of the Approved Document.
- Coursed brickwork or blockwork cavity walls should have at least 90 mm thick leaves, 50 mm cavity, and the combined width + 10 mm should be not less than the thickness determined by paragraph 1C6 above.
- Maximum spacing of cavity wall ties: 900 mm (horizontal) × 450 mm (vertical) spacings for 50–75 mm cavity; 750 mm (horizontal) × 450 mm (vertical) spacings for 76–100 mm cavity.
- Heights of buildings are related to slope of site and exposure.

Table 3.2   Solid external and compartment wall thicknesses

Maximum wall panel dimensions (m)	Minimum wall thickness (mm)
	190 for full height
	290 for one story, 190 for remainder
	290 for two storeys, 190 for remainder

- Wall panels require end restraint by way of a full-height buttressing wall, pier or chimney.
- Design criteria and limitations of openings, recesses, overhangs and chases are stated in the Approved Document.
- Wall panels require lateral restraint to restrict movement of the wall at right angles to its plane; consideration will therefore need to be given to strapping details at floor and roof levels.
- Reduced requirements apply to small buildings with single leaf walls, including the need for buttressing walls or piers at 3 m centres.

Section 1D: Height of chimneys

- Masonry used for the construction of chimneys to have a density of at least 1500 kg/m^3.
- Maximum height (at highest roof intersection and including any pot) to be at least 4.5 x W, where W is the least horizontal dimension.

Section 1E: Plain concrete strip foundations

The Approved Document provides rules for the design of foundations, including concrete mix, minimum widths and foundation dimensions. Table 3.3 gives minimum strip foundation widths.

Section 2: External wall cladding

The section gives guidance on the support and fixing of external wall cladding, and generally relates to heavier forms of cladding (e.g. stone or concrete), although some of the contents can be applied to curtain walling.

Table 3.3 Minimum strip foundation widths

Type/condition of subsoil	Minimum width (mm) for wall loading of					
	20	30	40	50	60	70 kN/ linear metre
Rock	- - - - - - - - - width of wall - - - - - - - - -					
Compact gravel or sand	250	300	400	500	600	650
Stiff clay or sandy clay	250	300	400	500	600	650
Firm clay or sandy clay	300	350	450	600	750	850
Loose sand, silty sand or clayey sand	400	600	- - - - to be calculated - - - -			
Soft silt, clay, sandy clay or silty clay	450	650	- - - - to be calculated - - - -			
Very soft silt, clay, sandy clay or silty clay	600	850	- - - - to be calculated - - - -			

To meet Requirement A1 the wall cladding should:

- safely sustain and transmit loadings to supporting structure;
- be securely fixed to give both vertical support and lateral restraint;
- make allowance for differential movement; and
- use durable materials, noting degree of accessibility.

Section 2 goes on to outline a technical approach, making reference to relevant British Standards, including BS 5080: Part 1: 1974 and Part 2: 1986, BS 5628: Part 3: 1985, BS 8200: 1985 and BS 8298: 1989.

Section 3: Recovering of roofs

The recovering of a roof is classified as a **material alteration** under Regulation 3. A procedure should then be followed comprising:

- a comparison between the proposed roof loading and the existing covering to establish if substantially higher or lighter loads will apply;
- an inspection of the existing roof structure to check whether it is capable of sustaining the increased loads and has adequate vertical restraint against wind uplift if a lighter covering and/or new underlay is proposed;
- the adoption of appropriate strengthening measures, for example the replacement of defective elements, provision of additional structural members and strapping provisions to include resistance to uplift.

Section 4: Codes, standards and references

This section lists various codes and standards that may be used for the structural design and construction of *all* buildings (Table 3.4).

With respect to foundation design particular attention should be paid to:

- the need to research conditions of ground instability, both known or recorded, and
- the availability, from the Department of the Environment, Transport and the Regions, of **reviews** containing regional reports of various geotechnical conditions.

Regulations 5 and 6 apply Requirements A1-A3 to certain buildings undergoing a material change of use. To appraise these buildings structurally, reference can be made to the following documents:

- BRE Digest 366: Structural appraisal of existing buildings for change of use (1991);

Table 3.4 Design codes and standards

Loading		BS 6399: Part 1: 1984 (dead and imposed loads) BS 6399: Part 3: 1988 (imposed roof loads) CP 3: Chapter V: Part 2: 1972 (wind loads)
Timber		BS 5268: Part 2: 1991 BS 5268: Part 3: 1985
Masonry		BS 5628: Part 1: 1978 BS5628: Part 3: 1985
Concrete		BS 8110: Part 1: 1985 BS 8110: Part 2: 1985 BS 8110: Part 3: 1985
Steel		BS 5950: Part 1: 1990 BS 5950: Part 2: 1992 BS 5950: Part 3: 1990 BS 5950: Part 4: 1982 BS 5950: Part 5: 1987 BS 449: Part 2: 1969
Aluminium		CP 118: 1969
Foundations		BS 8004: 1986

- The Institution of Structural Engineers Report *Appraisal of existing structures* (1980).

For information other guidance sources include:

- TRADA Approved Document Timber intermediate floors for dwellings;
- National House Building Council Standards;
- Zurich Municipal Builders Guidance Notes.

REQUIREMENT A3: DISPROPORTIONATE COLLAPSE

The building shall be constructed so that in the event of an accident the building will not suffer collapse to an extent disproportionate to the cause.

This requirement followed the Ronan Point disaster, in which an accidental explosion caused the partial collapse of a tower block containing flats.

Section 5: Reducing the sensitivity of the building to disproportionate collapse in the event of an accident

The application of the requirement is limited to a building having five or more storeys inclusive of basement storeys but *excluding* a storey within the roof space. This is of particular importance if the proposal is to add a new storey on top of an existing four-storey building.

To show compliance, a strategy is adopted that can be summarized as follows.

- Provide effective horizontal and vertical ties.
- If vertical tying is not possible, each member should be considered in turn to establish whether its position can be bridged, should it be removed.
- If neither vertical or horizontal tying is possible consider removal of each support member so as to establish the area at risk of collapse. This should not exceed 70 m^2 or 15% of storey area (see Diagram 25 of the Approved Document).
- Where the removal of a particular member gives rise to an excessive risk area it should be protected and therefore regarded as a **key element**.

BS 5628, BS 5950 and BS 8110 are referred to as an **alternative approach**.

Requirement A4 concerned the consequences of roof failure (or part thereof) and applied to public buildings, shops or shopping malls incorporating a clear span in excess of 9 m. Due to the existence of satisfactory control under Requirement A1 and support from recently published British Standards this requirement has been *deleted*. Because of this only Section 5, dealing with Requirement A3, should be referred to within the Approved Document.

The Approved Document concludes with Appendix A – Tables of sizes of timber floor, ceiling and roof members in single-family houses and a list of standards referred to.

CHAPTER 4

Approved Document B: Fire safety

INTRODUCTION

As a consequence of the review of the Building Regulations 1985 undertaken by the Department of the Environment, Approved Document B has changed significantly in a number of key areas, reflecting advances in technology and building form. These areas include the adoption of guidance rather than mandatory rules for means of escape design, a new requirement covering fire service access, and a lessening of standards concerning compartmentation and periods of fire resistance where the benefits of sprinkler systems are further recognized.

The 2000 edition supports the amended Part B Requirements which came into force on 1 July 2000. A number of significant changes have been incorporated within the guidance of the Approved Document.

The specific aims of requirements B1, B2, B3, B4 and B5 are highlighted within the general introduction to Approved Document B, and it is compliance with these different aspects of fire safety that must be considered by the designer:

- satisfactory means of warning and escape;
- restriction of fire spread over internal surfaces;
- sufficient stability of the building under fire load;
- sufficient fire separation within the building and between adjoining buildings;
- restriction of fire and smoke spread within concealed spaces;
- restriction of fire spread over the external envelope and from one building to another; and
- satisfactory access for fire appliances and firefighters up to and within the building in the saving of life.

Particular emphasis is placed on their close interrelationship, and where a variance is proposed of one or more of the requirements a trade-off could be possible so as to achieve an acceptable **fire safety package**.

Following on from this, the concept of **fire safety engineering** is introduced so as to achieve a *total* **fire safety package**. This is especially appropriate for large and complex building forms and existing buildings; smaller projects may also benefit from such an approach. The following specific factors should be assessed:

- risk of fire occurring in the first place;
- fire severity based on the fire load of the building (and its contents);
- structural resistance to fire and smoke spread; and
- potential risk to people in and about the building.

With these factors in mind a range of measures can be applied in varying degrees, including:

- adequate methods of fire prevention;
- swift warning of fire;
- control of smoke movement;
- utilization of active extinguishment methods;
- facilities to assist the fire service;
- staff training in fire safety and evacuation procedures; and
- utilization of continuing control under other legislation.

To quantify these measures a range of fire safety engineering techniques are available. These include zone modelling, computational fluid dynamics, virtual reality simulations, risk analysis, fire threat factors and statistical analysis. Further reference can be made to the BSI DD 240: 1997 *Fire safety engineering in buildings*.

OTHER LEGISLATION

In addition to the Building Regulations, consideration may need to be given to the application of other legislation that may impose fire safety requirements, including:

- the Fire Precautions Act 1971, where it will be necessary to establish if the premises require a fire certificate;
- the Fire Precautions (Workplace) Regulations 1997 (as amended);
- the Housing Act 1985 (as amended), for houses in multiple occupation;
- the Cinemas Act 1985, Gaming Act 1968, Licensing Act 1964 (as amended) Theatres Act 1968; and the Local Government (Miscellaneous Provisions) Act 1982, for licensed premises;
- the Safety of Sports Grounds Act 1975 (as amended);
- local Acts, where the local building control authority should be consulted.

One final point to bear in mind is that compliance with the Building Regulations is primarily for **life safety**, and offers only a reasonable standard of **property protection** in case of fire. It may be necessary to incorporate additional measures subject to clarification from the building insurers. Examples include the provision of fire walls, sprinkler protection and fire-resisting external walls, reference the LPC *Design guide for the fire protection of buildings*.

PROCEDURAL GUIDANCE

The increased scope of requirements contained in the 1992 edition of Approved Document B has influenced and changed the responsibilities of the enforcing authorities. For the majority of cases the building control body (approved inspector or local authority) will act as the Co-ordinating authority and have primary responsibility for enforcement concerning pre-occupation, with post-occupation enforcement undertaken by the fire authority.

Adequate and meaningful consultations between these parties and in turn with the designer and owner are therefore of the utmost importance. Because of this the Department of the Environment Transport and the Regions/Home Office/National Assembly for Wales have published a document entitled *Building Regulation and Fire Safety – Procedural Guidance*. The revision of the original guide takes into account the Fire Precautions (Workplace) Regulations and changes to the Approved Inspector Regulations. The aim of the document is to offer designers, developers and occupiers a procedural guide for all fire safety issues concerning buildings and building work. It should also act as a model for all enforcing authorities.

The guide gives a step-by-step process, which is summarized in Figure 4.1, highlighting the division of responsibilities.

A number of important items are raised within the guide, and are listed here for information:

- For a **designated or relevant use** the building control body *must* consult the fire authority.
- For a non-designated use the building control body *may* consult the fire authority.
- For premises subject to the licensing or registration (e.g. a concert hall or a residential care home) the building control body *should* consult with the fire authority.
- Other legislation and local Acts may also require consultation between the building control body and the fire authority.
- Where a fire risk assessment has been undertaken of the proposals, for the Workplace Fire Regulations (and/or the Construction (Health, Safety and Welfare) Regulations), it may prove useful to submit copies with the building regulation application.
- From the flowchart it can be seen that before the plans are passed to the fire authority they should show compliance with Part B.

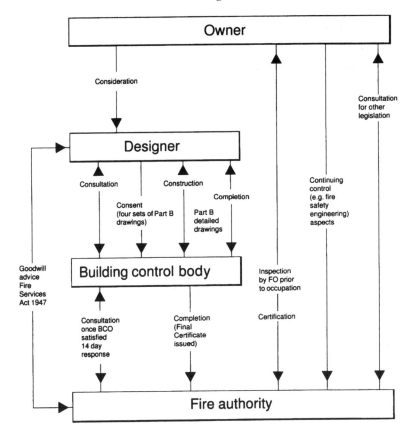

Figure 4.1 Procedural guidance flowchart.

- Where a building is controlled and approved under the Building Regulations with respect to means of escape the fire authority *shall* issue a fire certificate without requiring additional structural or other requirements relating to means of escape. This constraint is called the **statutory bar**.
- The Statutory Bar does not apply for matters not subject to building regulations (or not known at the time of approval) or in buildings covered by the Workplace Fire Regulations.
- The guidance suggests that *all* fire-related matters be indicated on the plans so that at the time of building occupation a fire certificate can be issued as soon as possible.
- Where differences of opinion occur between building control and the fire authority they should be dealt with between themselves up to the highest level; if deemed necessary the Department of the Environment, Transport and the Regions may be consulted.
- Note that the building regulations do not address the risk of fire

during construction, this is covered by the Construction (Health, Safety and Welfare) Regulations. Enforced by the fire authority for occupied buildings and the Health and Safety Executive when unoccupied.

PURPOSE GROUP

Before giving consideration to the guidance contained in the Approved Document the **purpose group(s)** of the proposal must first be established. A purpose group is a method of grading of occupancies based on assumed fire loadings. For this, reference should be made to Table 4.1. A building or compartment may contain more than one purpose group.

In the following situations the different or **ancillary use** should be regarded as a purpose group in its own right:

- a flat or dwelling;
- an area exceeding one fifth of the total building or compartment floor area where the building or compartment is over 280 m^2; or
- storage in a shop or commercial building (purpose group 4) exceeding one third of the total building or compartment floor area, where the building or compartment is over 280 m^2.

For buildings with more than one main use, which are not ancillary to each other, each use should be regarded as a purpose group in its own right. Offices over a ground-floor shop is one example. In large buildings where a complex mix of uses occurs consideration should be given to the risk of one use against the other(s). A shopping mall is an example where special measures may be necessary to offset the risk from one shop unit to another (see Section 12).

REQUIREMENT B1: MEANS OF ESCAPE

The building shall be designed and constructed so that there are appropriate provisions for the early warning of fire, and appropriate means of escape in case of fire from the building to a place of safety outside the building capable of being safely and effectively used at all material times.

The principle of providing people with a safe and effective means of escape from the building to a place of safety outside the building applies to all buildings with the exception of prisons provided under Section 33 of the Prisons Act 1952.

Table 4.1 Classification of purpose groups[1]

Title	Group	Purpose for which the building or compartment of a building is intended to be used
Residential[2] (dwellings)	1(a)	Flat or maisonette.
	1(b)	Dwelling house which contains a habitable storey with a floor level which is more than 4.5 m above ground level.
	1(c)	Dwelling house which does not contain a habitable storey with a floor level which is more than 4.5 m above ground level.
Residential (institutional)	2(a)	Hospital, home, school or other similar establishment used as living accommodation for, or for the treatment, care or maintenance of persons suffering from disabilities due to illness or old age or other physical or mental incapacity, or under the age of 5 years, or place of lawful detention where such persons sleep on the premises.
(Other)	2(b)	Hotel, boarding house, residential college, hall of residence, hostel and any other residential purpose not described above.
Office	3	Offices or premises used for the purpose of administration, clerical work (including writing, bookkeeping, sorting papers, filing, typing, duplicating, machine calculating, drawing and the editorial preparation of matter for publication, police and fire service work), handling money (including banking and building society work), and communications (including postal, telegraph and radio communications) or radio, television, film, audio or video recording, or performance [not open to the public] and their control.
Shop and commercial	4	Shops or premises used for a retail trade or business (including the sale to members of the public of food or drink for immediate consumption and retail by auction, self-selection and over-the-counter wholesale trading, the business of lending books or periodicals for gain and the business of a barber or hairdresser) and premises to which the public is invited to deliver or collect goods in connection with their hire, repair or other treatment, or (except in the case of repair of motor vehicles) where they themselves may carry out such repairs or other treatments.

Title	Group	Purpose for which the building or compartment of a building is intended to be used
Assembly and recreation	5	Place of assembly, entertainment or recreation; including bingo halls, broadcasting, recording and film studios open to the public, casinos, dance halls; entertainment, conference, exhibition and leisure centres; funfairs and amusement arcades; museums and art galleries; non-residential clubs, theatres, cinemas and concert halls; educational establishments, dancing schools, gymnasia, swimming pool buildings, riding schools, skating rinks, sports pavilions, sports stadia; law courts, churches and other buildings of worship, crematoria; libraries open to the public, non-residential day centres, clinics, health centres and surgeries; passenger stations and termini for air, rail, road or sea travel; public toilets, zoos and menageries.
Industrial	6	Factories and other premises used for manufacturing, altering, repairing, cleaning, washing, breaking up, adapting or processing any article, generating power, or slaughtering livestock.
Storage and other non-residential[3]	7(a)	Place for the storage or deposit of goods or materials [other than described under 7(b)] and any building not within any of the purpose groups 1–6.
	7(b)	Car parks designed to admit and accommodate only cars, motorcycles and passenger or light goods vehicles weighing no more than 2500 kg gross.

Notes:
[1] This table only applies to Part B.
[2] Includes any surgeries, consulting rooms, offices or other accommodation, not exceeding 50 m^2 in total, forming part of a dwelling and used by an occupant of the dwelling in a professional or business capacity.
[3] A detached garage not more than 40 m^2 in area is included in purpose group 1(c); as is a detached open carport of not more than 40 m^2, or a detached building which consists of a garage and open carport where neither the garage nor open carport exceeds 40 m^2 in area.

The **guidance** for the designer contained within the Approved Document is aimed at more straightforward projects and hence reference is made to a wide range of other guidance documents, including:

- British Standard 5588: Parts 0, 1, 4–11;
- Firecodes and Health Technical Memorandums 81, 88 etc.;
- *Draft guide to fire precautions in existing residential care premises;*

- *Guide to fire precautions in existing places of work that require a fire certificate: Factories, offices, shops and railway premises;*
- *Guide to Safety at Sports Grounds* (HMSO).

With regard to certain assembly buildings, e.g. theatres, lecture halls and stadia, problems arise with respect to fixed seating arrangements. The approved document acknowledges this point and makes direct reference to:

- BS 5588: Part 6: 1991 *Code of Practice for places of assembly,* and
- *Guide to safety at sports grounds* (HMSO),
- Reference can also be made to the *Guide to Fire Precautions in Existing Places of Entertainment* and *Like Premises* (HMSO).

For schools and other education buildings the fire safety objectives of the Department for Education and Employment's Constructional Standards can be achieved by following the Approved Document guidance, noting that *Building Bulletin 7* has been withdrawn.

Where individual shops are brought together to form a complex or shopping centre the principles for means of escape design vary, and the Approved Document recognizes this fact by referring the designer to:

- BS 5588: Part 10: 1991 Code of practice for enclosed shopping complexes, and
- Building Research Establishment Report 368 *Methodologies for smoke and heat ventilation.*

With respect to means of escape for disabled people Part M requires certain buildings (or parts) to have suitable access provisions. Therefore reasonable means of escape provisions should be made with reference to BS 5588: Part 8: 1988 *Code of practice for means of escape for disabled people*, although this may not necessarily involve structural measures.

B1 is split into six sections, which recognize the need for fire alarm systems and the separate components of the escape route including those particular to dwellings.

Section 1: Fire alarm and fire detection systems

Requirement B1 makes specific reference to the early warning of fire and the appropriate provisions for fire alarm systems. The guidance for dwellings has been expanded to include loft conversions and buildings other than dwellings, where previously the fire authority would control such installations under the Fire Precautions Act.

Suitable provisions for dwellings are:

Self-contained mains-operated smoke alarms (to BS 5446: Part 1) or a system in accordance with BS 5839: Part 1, to at least an L3 standard, or BS 5839: Part 6, to at least a Grade E type LD3 standard. For large houses, over 200 m^2, adopt Grade B type LD3 for under three storeys and an L2 standard where houses are over three storeys.

- For a loft conversion, in a one or two storey house, linked smoke alarms should be installed.
- For a flat or maisonette provide self-contained mains-operated smoke detection and alarms, including the minimum requirement for maisonettes of one unit per storey interconnected.
- For sheltered housing schemes, with a warden or supervisor, a connection to a central monitoring point should be incorporated.

Buildings other than dwellings should all have arrangements for detecting fire. The specification of the fire alarm/detection system will depend on the building type, purpose group and means of escape strategy.

- Small buildings may only require a shouted warning of "FIRE" that can be heard by all occupants of the premises.
- Manually operated sounders (handbells) or a simple manual call-point with a bell and power source may also prove suitable means to ensure compliance.

- Larger buildings will require the consideration of an electrically operated fire alarm system to BS 5839: Part 1. Type L for the protection of life, type P for property protection, type X for multi-occupancy buildings and type M, manual alarm system, which should be suitable for the majority of cases (purpose groups 3, 4, 5, 6, 7(a) and 7(b)), i.e. manual break-glass call-points and sufficient sounders.
- Shopping complexes and large assembly buildings may warrant a voice alarm, linked with the public address system, reference BS 5588: part 6 and part 10.
- Automatic fire detection and alarm systems should be provided for purpose groups 2(a) and 2(b), for example a Hotel would normally require a type L2 system protecting the escape routes and defined parts of the building.

Other circumstances may require automatic fire detection:

- protection of an inner room situation
- compensation for a trade-off where variance is proposed,
- to operate pressurisation systems or automatic door release mechanisms,
- to protect unoccupied areas which could prejudice escape from occupied areas, and
- for phased evacuation of a building.

Fire detection and alarm systems must be properly designed, installed and maintained. The installation and commissioning certification should be issued to the building control body and fire authority.

Section 2: Dwelling houses

The provisions for one- and two-storey houses are straightforward but important:

- All habitable rooms (excluding kitchens) in the upper storey(s), served by a single stair, should have a window (or external door)at least 0.33 m^2, 450 mm high, 450 mm wide and not more than 1100 mm above floor level.
- All habitable rooms (excluding kitchens) should open directly onto an entrance hall or other suitable exit or should have a window or door as above.
- A sleeping gallery, not more than 4.5 m above ground level, will require an alternative exit if the distance from any part to the head of the access stair exceeds 7.5 m. Also a distance of 3 m should not be exceeded from the foot of the access stair to the room exit and cooking facilities should be enclosed or remote from the access stair.
- Early fire warning, see Section 1.

- Habitable rooms within basements should have alternative escape routes.

Note that a room is termed an **inner room** where its only escape route is via another room or **access room**. Only acceptable for a kitchen, laundry or utility room, dressing room, bath/shower room, wc, other room on floor up to 4.5 m and a sleeping gallery. An open-plan house arrangement with a stair open to a ground-floor room would give rise to this situation. Guidance for the installation of smoke alarms is given in Section 1, including the minimum requirement of one unit per storey, interconnected where there is more than one.

For houses that have one or more floors above 4.5 m, measured from ground level, more stringent provisions apply. This is due to the greater risk associated with escape via high-level windows. Provisions for three-storey houses, in addition to the above, are:

- the utilization of a **protected stairway** giving directly to a final exit (or via two final exits separated by fire-resisting construction); or
- the separation of the top storey by fire-resisting construction and the provision of an **alternative escape route** leading to its own final exit.

A protected stairway is a minimum half-hour fire-resisting enclosure containing the staircase, having self-closing fire doors, designated FD 20, to all rooms (except a bathroom or WC) and discharging to a final exit.

For loft conversions to existing two-storey dwelling houses, not exceeding 50 m² in floor area, containing no more than two habitable rooms and where work is not taken above the existing ridge line, reference can be made to paragraphs 2.18-2.25 of the Approved Document, which relaxes the above provisions. The main provisions are illustrated in Figure 4.2.

For houses of four storeys and above an alternative escape route should be provided to all floor levels over 7.5m. Reference should be made to BS 5588: Part 1: 1990 *Code of practice for residential buildings*.

Section 3: Flats and maisonettes

As with housing, few provisions are necessary for basement, ground and first floor storeys, i.e. inner rooms and smoke alarms. It is only with the increase in height that the degree of risk increases, calling into play further detailed provisions. The inherent compartmentation between flats (see also Requirement B3) leads to a low probability of fire spreading from the dwelling of origin. Hence the complete evacuation of the building is not always necessary, and residents can therefore remain within their homes. Satisfactory means of escape provisions must still be provided, however, and these should allow

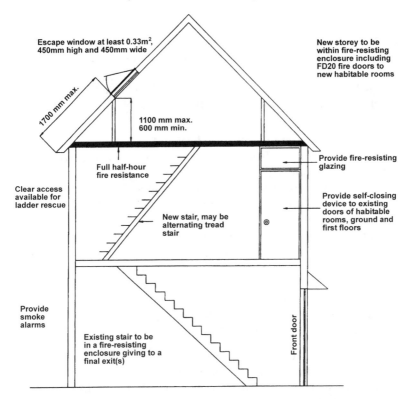

Escape window at least 0.33m², 450mm high and 450mm wide

New storey to be within fire-resisting enclosure including FD20 fire doors to new habitable rooms

1700 mm max.

1100 mm max. 600 mm min.

Full half-hour fire resistance

Provide fire-resisting glazing

Clear access available for ladder rescue

Provide self-closing device to existing doors of habitable rooms, ground and first floors

New stair, may be alternating tread stair

Provide smoke alarms

Existing stair to be in a fire-resisting enclosure giving to a final exit(s)

Front door

Figure 4.2 Provisions for loft conversion.

the safe evacuation from the building to a place of ultimate safety, i.e. the outside air at ground level. In the design of such provisions specific consideration needs to be given to the spread of smoke, prevention of fire and fire spread within the common parts of the building, and the fact that reliance should not be placed on external rescue.

In broad terms the Approved Document guidance reflects the principles contained within BS 5588: Part 1: 1990, and indeed reference is made to it for other less common layouts. The Approved Document guidance is also applicable to **houses in multiple occupation** and **sheltered housing schemes**.

Escape routes within each dwelling and escape from the dwelling itself, via any common corridor or stair, to the final exit, represent the two components for satisfactory means of escape design, which can be summarized as follows.

Escape within dwelling

For one- and two-storey flats and maisonettes, where a floor level is not more than 4.5 m above ground level, apply the guidance in Section 2.

For floors more than 4.5 m above ground level adopt (a), (b) or (c) for flats and (d) or (e) for maisonettes (without their own external entrance):

(a) Provide a **protected entrance** hall, where the maximum **travel distance** within it from a door to any habitable room to the entrance door does not exceed 9 m, *or*

(b) limit the maximum overall travel distance, from any point in the flat, to the entrance door to 9 m, with the cooking facilities kept remote from the escape route, *or*

(c) provide an alternative exit from the bedroom accommodation, which itself should be separated from the living accommodation by half-hour fire-resisting construction.

(d) Provide **alternative exits** to each habitable room not located on the entrance floor level, *or*

(e) an **alternative exit** to the floor level (not entrance floor level) utilizing a **protected landing/hall**.

Note that an alternative exit should be remote from the main entrance door to the dwelling and lead to a final exit or common stair via a suitably protected access corridor, common balcony, internal private stair, external stair, common stair or a route over a flat roof (cross-reference Approved Document K).

- Provide *within* each flat or maisonette early fire warning, see Section 1, including the minimum requirement for maisonettes of one unit per storey interconnected.

Escape within common areas

- Each dwelling, on a floor above 4.5 m from the ground level, should have an alternative escape route from the entrance door so that a person can turn and escape away from the fire. This is not necessary where each dwelling has access to a single common stair which is via a **protected lobby** or corridor and the entrance door is within a maximum travel distance of 7.5 m to the stair enclosure (this distance is also applicable within the dead end portion of a common corridor; see Figure 4.3).
- Where an alternative escape route is available from the entrance door (i.e. to more than one common stair) then the travel distance within the common corridor to a common stair should not exceed

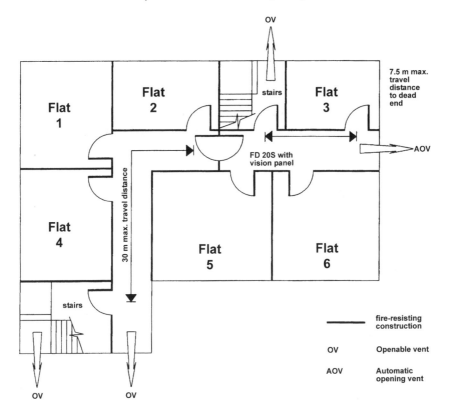

Figure 4.3 Provisions for flats served by two common stairs.

30 m. The corridor in this case should be subdivided by a self-closing fire door (FD 20S designation incorporating smoke seals). See Figure 4.3.

- For small single-stair buildings modified guidance applies, where a maximum travel distance of 4.5 m from the entrance door to the stair enclosure is acceptable or no travel distance limit where each storey only contains two dwellings (see Diagram 14 of the Approved Document) subject to:

    (a) the top floor not exceeding 11 m (from ground level), and
    (b) not more than three storeys above ground level, and
    (c) the stair not to connect to a covered car park and
    (d) the stair not to serve ancillary accommodation unless separated with ventilated protected lobby/corridor and no dwelling is located on ancillary accommodation floor.

- Even with the inherent compartmentation provided within flat and maisonette developments consideration still needs to be given to smoke dispersal within the common escape routes. This is achieved by **openable vents** (OV) (minimum 1.0 m^2 free area) or **automatic opening vents** (AOV) (minimum 1.5 m^2 free area).

- Cross-corridor fire doors (where provided) and OVs or AOVs may be omitted where a pressurization system in accordance with BS 5588: Part 4 is employed.
- The minimum width for a common stair should be 1.1 m if used as a firefighting stair, otherwise the width for everyday use is acceptable.
- All common stairs (including any passageway to the final exit) should be constructed as protected stairs and enclosed with fire-resisting construction.
- Two adjoining protected stairways (or exit passageways) should be separated by imperforate fire-resisting construction.
- It is important that a protected stairway contains no potential fire risk; the only allowable exceptions are a lift well or electrical meter(s).
- The stair in a single-stair building should not descend to a basement storey or serve any covered car park, boiler room, fuel storage space or any other ancillary accommodation of similar high fire risk. Other stairs may communicate with these areas subject to the provision of a ventilated protected lobby/corridor at each level.
- Dwellings on not more than three storeys within mixed-use buildings may communicate with other occupancies, where the stairs are separated by protected lobbies. Flats above three storeys should be provided with independent access unless, the dwelling is ancillary, has an independent alternative escape route, linked fire alarm system and the stairs separated by protected lobbies.

Section 4: Horizontal escape for buildings other than dwellings

As with Section 2 the means of escape design should allow persons to turn their back on a fire and evacuate the building safely. This particular section, applying to all other buildings, considers the portion of the escape route from any point on the floor to the storey exit. This is illustrated in Figure 4.4, where the Boardroom, point A, is regarded as an inner room from which there is only one route of escape (see also 45°rule in Diagram 15 of the Approved Document). However, alternative escape routes are available from point B.

With the above in mind the first task is to establish the population of any room, tier or storey. If this is not known from the design brief or actual data (average occupant density at a peak trading time of the year) then reference can be made to Table 4.2.

Once the population of any room, tier or storey has been established the minimum number of escape routes/storey exits from that floor level can be found from Table 4.3.

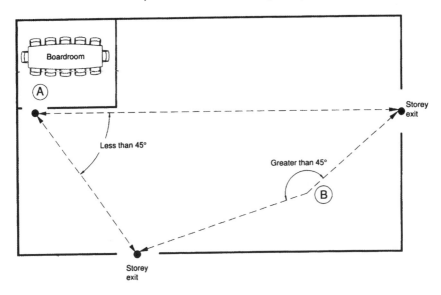

Figure 4.4 Alternative escape routes.

Reference can now be made to Table 4.4, which brings together limitations on **travel distance**. It relates purpose groups to maximum travel distances for escape in one direction only or where escape is available in more than one direction.

Consideration now needs to be given to the adequate width of escape routes and storey exits based on the following:

- 750 mm for up to 50 persons (530 mm between fixed storage racking);
- 850 mm for up to 110 persons;
- 1050 mm for up to 220 persons and 5 mm per person thereafter.

It is important to take into account the possibility of fire and/or smoke obstructing the use of one of the available exits (Figure 4.5). Therefore where two or more exits exist, the largest one must be discounted so as to determine the adequate width of the remainder. Reference should also be made to Approved Document M concerning adequate widths for disabled access.

The remaining provisions outlined within the Approved Document with respect to horizontal escape can be summarized and illustrated as follows.

- To protect persons in an **inner** room against the occurrence of fire in the **access** room the enclosing walls of the inner room should be stopped a minimum 500 mm short of the ceiling, *or* a door or wall vision panel provided, minimum 0.1 m^2 in area, *or* a suitable

Table 4.2   Floor space factors

Type of accommodation[1,6]	Floor space factor $(m^2/person)$
1. Standing spectator areas, bars without seating and similar refreshment areas.	0.3
2. Amusement arcade, assembly hall (including a general purpose place of assembly), bingo hall, dance floor or hall, club, crush hall, venue for pop concert and similar events.	0.5
3. Concourse, queuing area or shopping mall[2].	0.7
4. Committee room, common room, conference room, dining room, licensed betting office (public area), lounge or bar (other than in 1. above), meeting room, reading room, restaurant, staff room, waiting room[3].	1.0
5. Exhibition hall or Studio (film, radio, television, recording).	1.5
6. Shop sales area[4], skating rink.	2.0
7. Art gallery, dormitory, factory production area, Museum or workshop.	5.0
8. Office.	6.0
9. Kitchen, library, shop sales area[5].	7.0
10. Bedroom or study bedroom.	8.0
11. Bedsitting room, billiards or snooker room or hall.	10.0
12. Storage and warehousing.	30.0
13. Car park.	Two persons per parking space

Notes:

[1] Where accommodation is not directly covered by the descriptions given, a reasonable value based on a similar use may be selected.

[2] Refer to section 4 of BS 5588: Part 10 for detailed guidance on the calculation of occupancy in common public areas in shopping complexes.

[3] Alternatively the occupant capacity may be taken as the number of fixed seats provided, if the occupants will normally be seated.

[4] Shops excluding those under item 9, but including supermarkets and department stores (main sales areas), shops for personal services such as hairdressing and shops for the delivery or collection of goods for cleaning, repair or other treatment or for members of the public themselves carrying out such cleaning, repair or other treatment.

[5] Shops (excluding those in covered shopping complexes and including department stores) trading predominantly in furniture, floor coverings, cycles, prams, large domestic appliances or other bulky goods, or trading on a wholesale self-selection basis (cash and carry).

[6] If there is to be mixed use, the most onerous factor(s) should be applied.

Table 4.3 Minimum escape routes

Maximum number of persons	*Minimum number of escape routes/exits*
60	1
600	2
Above 600	3

Note: A single escape route or storey exit may be acceptable where the travel distance in one direction to a storey exit can be complied with and the established population for the area or storey (excluding a storey used for in-patient care in a hospital) does not exceed 60 persons (or 30 for an institutional use, purpose group 2a).

automatic fire detection and alarm system so as to alert the occupants of the inner room. This arrangement is only acceptable under certain conditions:

(a) the inner room population should not exceed 60 persons (30 for purpose group 2a);
(b) the inner room should not be a bedroom;
(c) the escape route should pass through no more than one access room;
(d) the appropriate one-direction travel distance should be complied with; and
(e) the access room should be under the control of the same occupier and not be a place of special fire hazard (e.g. a switchgear or boiler room).

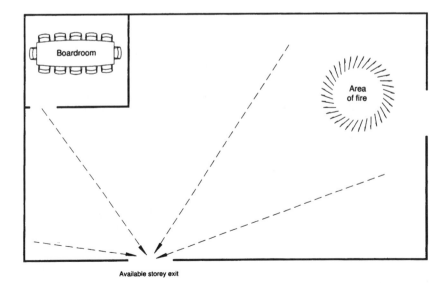

Figure 4.5 Escape widths.

Table 4.4 Limitations on travel distance

Purpose group	Use of the premises or part of the premises	Maximum travel distance[1] where travel is possible in:	
		One direction only (m)	More than one direction (m)
2(a)	Institutional[2]	9	18
2(b)	Other residential:		
	(a) in bedrooms[3]	9	18
	(b) in bedroom corridors	9	35
	(c) elsewhere	18	35
3	Office	18	45
4	Shop and commercial[4]	18	45
5	Assembly and Recreation;		
	(a) buildings primarily for the handicapped except schools	9	18
	(b) schools	18	45
	(c) areas with seating in rows	15	32
	(d) elsewhere	18	45
6	Industrial[5]	25	45
7	Storage and other non-residential[5]	25	45
2–7	Place of special fire hazard[6]	9[3]	18[3]
2–7	Plant room or roof top plant:		
	(a) distance within the room	9	35
	(b) escape route not in open air (overall travel distance)	18	45
	(c) escape route in open air (overall travel distance)	60	100

Notes:

[1] The dimensions in the Table are travel distances. If the internal layout of partitions, fittings etc. is not known when plans are deposited, direct distances may be used for assessment. The direct distance is taken as 2/3rds of the travel distance.

[2] If provision for means of escape is being made in a hospital or other health care building by following the detailed guidance in the relevant part of the Department of Health 'Firecode', the recommendations about travel distances in the appropriate 'Firecode' document should be followed.

[3] Maximum part of travel distance within the room.

[4] Maximum travel distances within shopping malls are given in BS 5588: Part 10: 1991. Guidance on associated smoke control measures is given in a BRE report 368.

[5] In industrial buildings the appropriate travel distance depends on the level of fire risk associated with the processes and materials being used. Control over the use of industrial buildings is exercised through the Fire Precautions Act. Attention is drawn to the guidance issued by the Home Office Guide to fire precautions in existing places of work that require a fire certificate, Factories, Offices, Shops and Railway premises. The dimensions given above assume that the premises will be of 'normal' fire risk, as described in the Home Office guidance. If the building is high risk, as assessed against the criteria in the Home Office guidance, then lesser distance of 12 m in one direction and 25 m in more than one direction, would apply.

[6] Places of special fire hazard are listed in the definitions in Appendix E.

- Storey exits into a central core arrangement need to be adequately separated to retain protected alternative escape routes (see Diagram 17 of the Approved Document).
- Travel via one stairway enclosure to reach another and the use of a stairway as a main circulation route at the same level should both be avoided, unless doors fitted with automatic release mechanisms.
- An ancillary area used for the consumption of food and/or drink by customers should have at least two direct escape routes (avoiding areas of higher fire hazard).
- Escape routes from different **occupancies** (i.e. separate ownerships or tenancies of different organizations) within the same storey should not pass via each other, and any common corridor or circulation space should be constructed as a protected route unless a suitable automatic alarm and detection system is installed throughout the storey.
- Corridors serving bedrooms, corridors common to different occupancies and dead-end corridors should be constructed as **protected corridors** (see Figure 4.6).
- Other corridors forming an escape route should have doors to room openings and have partitions taken to floor or ceiling soffit.
- Where corridors exceed 12 m in length and connect two alternative storey exits the corridor should be subdivided at a mid - position with minimum FD 20S self-closing fire doors (see Figure 4.6).
- To protect dead-end lengths of corridor exceeding 4.5 m (and not protected by a pressurization system to BS 5588: Part 4) they should

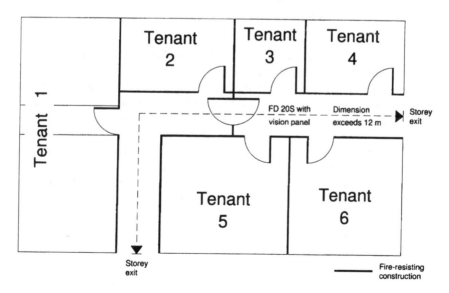

Figure 4.6 Protected corridor.

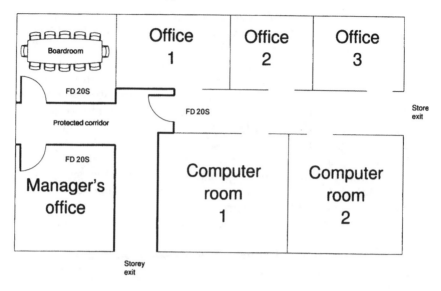

Figure 4.7   Dead-end corridor.

be separated by self-closing fire doors and fire-resisting construction from any corridor offering two routes of escape or continuation of a route past a storey exit so as to reach another (see Figure 4.7).

- The rules for the protection of external escape routes and those over flat roofs are the same as outlined in Section 3 for dwellings, of the Approved Document. Reference should also be made to Approved Document K – Stairs, ramps and guards.

Reference has already been made to Department of Health 'Firecode' documents, which may also be used for non-NHS premises. One of the fundamental principles that is applied to hospitals and other residential care premises is the utilization of **progressive horizontal evacuation**. This avoids the complete evacuation of the patients or residents from the building or any part of it, allowing escape horizontally into an adjoining compartment or place of relative safety. The door(s) in this line of compartmentation can be regarded as a storey exit, where the travel distance limitations should comply with Table 4.4. The Approved Document verifies that this principle may be of value in some other residential buildings, e.g. a residential care home for elderly mentally frail people, and guidance is offered in this regard for the designer, as follows:

- Each upper storey, used for in-patient care, should be subdivided into a minimum of two compartments.
- The adjoining compartment floor area should be sufficient to accom-

Table 4.5 Minimum stair widths

Situation	Maximum number of persons	Minimum stair width (mm)
Institutional building (excluding staff use)	150	1000
Assembly building (exceeding 100 m²)	220	1100
Any other building (occupancy above 50)	Over 220	See Table 4.6 or 4.7
Any other stair	50	800 mm

modate the combined population of the compartment at risk and the adjoining compartment itself.

- Each compartment should have a minimum of two independent escape routes. A route into an adjoining compartment may be via a further, or third compartment, subject to this having independent storey exits to the other two compartments.

Section 5: Vertical escape for buildings other than dwellings

To evacuate the population of multi-storey buildings from each floor level down to safety at ground level a sufficient number of escape stairs of adequate width are required. The number of stairs is determined by:

- travel distance limitations (Table 4.4);
- the need for separate stairs to serve assembly and recreation or residential purpose groups within a mixed use building;

Table 4.6 Capacity of a stair for basements and simultaneous evacuation of a building

No. of floors served	Maximum number of persons for stair width:								
	1000 mm	1100 mm	1200 mm	1300 mm	1400 mm	1500 mm	1600 mm	1700 mm	1800 mm
1	150	220	240	260	280	300	320	340	360
2	190	260	285	310	335	360	385	410	435
3	230	300	330	360	390	420	450	480	510
4	270	340	375	410	445	480	515	550	585
5	310	380	420	460	500	540	580	620	660
6	350	420	465	510	555	600	645	690	735
7	390	460	510	560	610	660	710	760	810
8	430	500	555	610	665	720	775	830	885
9	470	540	600	660	720	780	840	900	960
10	510	580	645	710	775	840	905	970	1035

Table 4.7 Minimum aggregate width of stairs designed for phased evacuation

Maximum no. of persons per storey	Stair width (mm)
100	1000
120	1100
130	1200
140	1300
150	1400
160	1500
170	1600
180	1700
190	1800

- whether a single stair is acceptable for floor levels not more than 11 m above ground level and where a single escape route at each floor level is acceptable (including a basement level); and
- adequate escape width for the established population likely to use the stairs in an emergency subject to the **minimum** values listed in Table 4.5.

To calculate minimum stair widths two strategies exist, depending on the most appropriate method of evacuating the building or part of the building.

The **simultaneous evacuation** of the building must be used for basements, buildings containing open spatial planning arrangements, other residential (e.g. hotel), and assembly and recreation buildings where minimum aggregate stair widths are calculated by either:

- reference to Table 4.6 for buildings up to 10 storeys; *or*
- use of the formula $P = 200w + 50(w - 0.3)(n - 1)$, where P = number of persons accommodated, w = width in metres, and n = number of storeys.

Phased evacuation is the second strategy option where the fire floor (i.e. floor of fire origin) and the floor above are evacuated first and thereafter at two floors at a time. This method allows less disruption to occupants of the building and gives stairs of reduced width due to the limited number of persons making their escape at any one time. A number of conditions apply:

- stairs to be approached via protected lobbies/corridors (except a top storey containing plant rooms);
- each floor to be constructed as a compartment floor;
- building to be protected throughout with a sprinkler system (to BS 5306: Part 2, including life safety provisions) if a floor is above 30 m, but this does not apply to a building containing flats;

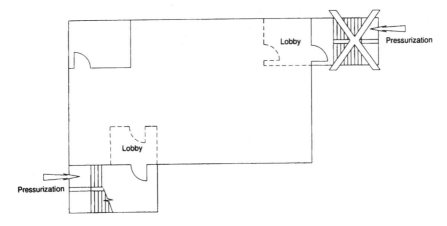

Figure 4.8   Provisions for stairways.

- an automatic alarm and detection system to be installed to at least L3 standard, BS 5839: Part 1, and an internal speech communication system to allow conversation links from fire brigade access level to fire warden at each floor level.

Phased evacuation may be used in high buildings (over 30 m) and large complex buildings. Minimum widths are calculated by either:

- reference to Table 4.7; *or*
- use of the formula [(P x 10) – 100] mm where P = number of persons accommodated on the most heavily occupied storey.

Both methods assume that one stair in a multi-stair building will not be available in an emergency, owing to its being smoke-logged or otherwise affected by fire. Each stair should therefore be **discounted** in turn so as to verify that the remainder have the necessary capacity. This discounting rule does not, however, apply where all escape stairs are approached at each floor level (except the top storey) via a protected lobby, *or* all the stairways are protected by a smoke control pressurization system in accordance with BS 5588: Part 4 (Figure 4.8). Note that a storey exit still needs to be discounted. In addition, a stair in excess of 30 m should not exceed 1400 mm in width or alternatively be provided with a central handrail for an 1800 mm minimum width.

The protection of escape stairs, including any passageway leading to the final exit, should follow established guidelines, which can be summarized as follows.

- Provide a protected enclosure to required period of fire resistance, unless an **accommodation stair** (i.e. not a protected stair) can form part of the internal escape route.

- A protected lobby or protected corridor is deemed necessary for: single-stair buildings (with more than one floor above or below the ground storey); buildings designed for phased evacuation; buildings where the stair serves a storey above 18 m; a fire fighting shaft); and where a stair needs to be separated from a special fire hazard (smoke ventilation by way of 0.4 m^2 permanent vent or mechanical smoke control required).
- Adjoining protected stairs should be separated by imperforate construction.
- It is particularly important to restrict potential sources of fire within protected stairs, especially those serving single-stair buildings. The allowable limitations are sanitary accommodation or washrooms (incorporating gas water heaters and sanitary towel incinerators only), a lift well (unless it is a firefighting stair), a reception desk or enquiry office up to 10 m^2 in area (or equivalent fire risk) but not in a single-stair building, and finally a cupboard enclosed with fire-resisting construction, again not allowable in a single-stair building.
- The protection for external walls of protected stairs and external stairs should follow the guidance in Section 6, for basement stairs refer to Section 3.

Section 6: General provisions for buildings other than dwelling houses

This section deals with the construction and protection of escape routes and service installations. It is important that it should be read in conjunction with Sections 3–5.

To safeguard escape routes during the evacuation of a building they should be protected to a minimum 30 min standard of fire resistance. Consideration also needs to be given to Requirements B3, B5 and other sections of B1 where greater periods of fire resistance may be needed (see Table 4.9). Where glazed elements are incorporated within fire-resisting construction, including doors, they should offer the required period of fire resistance in terms of both integrity and insulation. Where the glazing does not offer the insulation criteria, e.g. georgian wired glass, then strict limits apply to areas of use, as outlined in Table A4 of the Approved Document. This is because fire presents a radiation hazard for persons trying to escape past glazing with no insulation properties: hence the use of an 1100 mm minimum height (in certain circumstances) above floor level to allow escape under this glazing. Note that glazing restrictions also apply to firefighting shafts: see clause 9 of BS 5588: Part 5, and additional guidance on the safety of glazing is given in Approved Document N – Materials and protection.

Escape routes should also be constructed to offer a minimum 2 m headroom with no projections below this height with the exception of

door frames. The flooring chosen for all escape routes should be safe in use and minimize slipperiness when wet.

Doors on escape routes represent one of the most important aspects of means of escape design, in terms of both fire resistance and mode of operation. The following is a checklist of the items to consider, together with the contents of Table B1 of the Approved Document.

- Fire doors need to offer the required period of fire resistance and satisfy the test criteria outlined in Appendix B and Table B1 of the Approved Document.
- Generally, all fire doors forming part of an enclosure to common escape routes, protected corridors, lobbies and stairways need to prevent smoke leakage at ambient temperatures, denoted by the suffix (S) in the fire door designation.
- Hardware and ironmongery used on fire doors should not impair performance. Further guidance can be found in *Code of practice for hardware essential to the optimum performance of fire-resisting timber doorsets* published by the Association of Builders' Hardware Manufacturers.
- All doors on escape routes should not be fitted with a lock, latch or bolt. Simple fastenings operated from the escape side of the door may be used where only one mechanism needs to be manipulated without the use of a key. The use of 'panic bolts' or similar devices, in assembly or commercial uses for example, is allowable. Additional security locks may be acceptable, when the building is empty, subject to suitable management control.
- Doors should be hung to open in the direction of escape and must be so hung where the population served exceeds 50 persons.
- The swing of a door should be at least 90° and be clear of any change of floor level (except a threshold or single step), not reduce the effective width of a landing or corridor and incorporate a vision panel if hung to swing both ways.
- Revolving and automatic doors or turnstiles used for escape purposes should fail safe, in the open position, in an emergency. If this is not possible adjoining swing doors of sufficient width should be provided.

Turning to the construction of escape stairs, these should utilize **materials of limited combustibility** in the following locations:

- a single-stair building (excluding purpose group 1a and 3 up to three storeys);
- a basement storey (excluding a private stair within a maisonette);
- a storey more than 18 m above ground or access level;
- an external stair serving a storey more than 6 m above ground level;
- a firefighting stair.

Note that combustible materials may be added to the top surface of these stairs (excluding firefighting stairs).

The escape route may utilize helical and spiral stairs, subject to accordance with BS 5395 *Stairs, ladders and walkways*: Part 2, or a fixed ladder where it is not practical to provide a conventional stair, e.g. to serve a plant room or tank room. Fixed ladders should not be used by members of the public or be constructed of combustible materials. Ramps forming part of an escape route should comply with Approved Document M, and the pitch of a floor or tier slope, e.g. within a theatre, should not exceed 35°. Reference should also be made to Approved Document K – Stairs, ramps and guards.

The remaining general provisions can be summarized as follows.

- In certain projecting or recessed protected stairway locations the safe use of the stair may be threatened by fire emanating from adjoining accommodation. In these cases the adjoining external wall construction within 1.8 m should be fire-resisting.
- External walling, doors and windows within a 1.8 m zone (9 m below stair and 1.1 m above top landing) should offer a minimum half-hour period of fire resistance where adjacent to an external escape stair or route.
- Final exits should have a minimum width of the escape routes they serve; should be sited to achieve swift evacuation of the building and to avoid further threat from fire and smoke; should offer direct access to a street, passageway, walkway or open space; should be well defined, and guarded where necessary.
- Each doorway, exit and route provided for means of escape, except in dwellings and for exits in normal use, should be provided with suitably located and conspicuous fire exit signage in accordance with BS 5499: Part 1: 1984 or Health and Safety (Safety Signs and Signals) Regulations 1996, e.g. European running man pictogram. Note that additional signage and notices may be necessary to comply with other legislation.
- Protected power circuits should ensure a continuing power supply, where necessary, in the event of fire; see CWZ classification for cabling in accordance with BS 6387.
- Passenger lifts are not usually used for means of escape purposes. However, **evacuation lifts**, suitably sited and protected, can be utilized for the evacuation of disabled persons. For specific guidance reference should be made to BS 5588: Part 8 (and Part 5).
- Lift installations can influence escape routes and hence:

    (a) lift wells should be contained within the protected stair enclosure or enclosed with fire-resisting construction if means of escape is prejudiced;

(b) lift wells serving different compartments should be regarded and constructed as protected shafts;

(c) wall climber lifts may penetrate a smoke reservoir, e.g. within an atrium, where its integrity should be preserved;

(d) in basements and enclosed car parks and where the lift serves areas of high fire risk *and* dwellings then the lift should be approached via a protected lobby (or corridor);

(e) a lift in a single-stair building or a building where a protected stair enclosure terminates at ground level should not descend down to basement level(s);

(f) lift machine rooms should be sited above the lift well and outside the enclosure of a single-stair building.

- The design of any mechanical ventilation or air conditioning system should either close down, or direct potentially smoke-laden air away from any escape route, in the event of fire. Specific reference should be made to BS 5588: Part 9: 1989 (BS 5588: Part 6 and BS 5720).

- Storage chambers, chutes and hoppers used for refuse disposal should be in accordance with BS 5906. Note the need to separate refuse chutes and storage rooms within fire-resisting construction and access provided direct from the external air or via a permanently ventilated (0.2 m^2) lobby.

- Walk-in store rooms in shops, which are fully enclosed, should be separated from retail areas with fire resisting construction if they

Table 4.8   Provisions for escape lighting

Purpose group	Areas requiring escape lighting
Residential	All common escape routes (including external), except two storey flats.
Office, shop and commercial (where public not admitted), industrial, storage, other non-residential	Underground or windowless. accommodation, stairways in a central core or serving a floor above 18 m, internal corridor exceeding 30 m and open plan offices exceeding 60 m^2.
Shop and commercial and car parks (where public admitted)	All escape routes (including external), except small shops (not restaurant or bar) up to three storeys and maximum 280 m^2 sales floor area.
Assembly and recreation	All escape routes (including external), and accommodation except where open on one side (daylight hours) and schools (during normal hours).
Any purpose group	Toilets over 8 m^2 (and under it windowless), electrical generator rooms, switch/battery rooms for emergency lighting system and emergency control room.

could prejudice escape routes, unless automatic detection or sprinklers are fitted.

- All escape routes should be provided with adequate artificial lighting *and* **escape lighting** in accordance with BS 5266: Part 1: 1988, or CP 1007 for cinemas, to the areas listed in Table 4.8.

REQUIREMENT B2: INTERNAL FIRE SPREAD (LININGS)

1. To inhibit the spread of fire within the building, the internal linings shall:
   (a) adequately resist the spread of flame over their surfaces; and
   (b) have, if ignited, a rate of heat release which is reasonable in the circumstances.
2. In this paragraph 'internal linings' mean the materials lining any partition, wall, ceiling or other internal structure.

Section 7: Wall and ceiling linings

Wall and ceiling linings are not normally the source of fire, although fire spread and rate of fire growth across these surfaces can be crucial. This is especially so within escape routes and circulation areas, where the rapid spread of fire could prevent persons making their escape. The requirement seeks the use of lining materials that have low rates of surface spread of flame and low rates of heat release. It does not, however, address the problem of fumes and smoke generated by such linings. Compliance should be shown with the classifications indicated in Figure 4.9, depending on location. These classifications are based on fire tests in BS 476: Parts 6 and 7, where Class 1 is the highest, although Class 0 offers the best performance, but is not identified in any British Standard test.

Fire spread across the upper surfaces of floors and stairs is not regarded as a problem in the early stages of a fire and is not therefore controlled. The provision, location and specification of furniture and fittings is also not controlled since the continuing control of the building and its contents is not possible under building regulations. For certain premises that require a licence or fire certificate, control can be exerted by the fire authority.

The definition of **walls** *includes* the surface of glazing (except door glazing) and any ceiling sloping at more than 70° to the horizontal, but *excludes* doors and doorframes, window and other glazed frames, architraves, cover moulds, picture rails, skirtings and similar narrow

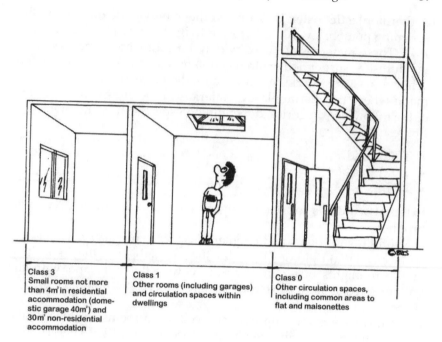

Figure 4.9. Classification of linings.

Note: A room is taken as an enclosed space that is not used for circulation only, and would include not just conventional rooms but spaces ranging in size from cupboards up to warehouses and auditoria.

members, fireplace surrounds, mantelshelves and fitted furniture. The definition for **ceilings** *includes* the surface of glazing, wall sloping at 70° or less from the horizontal but *excludes* trapdoors and frames, window, rooflight and other glazed frames, architraves, cover moulds, picture rails and similar narrow members.

A lower class may be adopted for walls in rooms, minimum Class 3, where the wall areas in question do not exceed half of the room floor area subject to a 20 m² maximum for residential buildings and 60 m² for non-residential buildings. A variation is also applicable with regard to rooflights where a lower class may be used, minimum Class 3, subject to the limitations contained in Tables 11 and 18 of the Approved Document.

Special applications are highlighted:

- Air supported structures should comply with BS 6661;
- Other flexible membrane covering a structure, reference Appendix A of BS 7157;
- PTFE-based materials, reference BRE Report 274.

A **thermoplastic material** is any synthetic polymeric material with a softening point below 200°C as tested to BS 2782: Part 1: Method 120A: 1976. These materials are used widely for rooflights, windows, lighting diffusers, suspended or stretched-skin ceilings. A material may not achieve a Class 0-3 designation, and therefore concessions can be made where the thermoplastic material can be classified under one of the following categories, depending on specification and method of fire test:

- TP(a) rigid;
- TP(a) flexible;
- TP(b).

External windows to rooms only may be glazed with a TP(a) rigid product.

Rooflights to rooms and circulation spaces only may have a lower surface of TP(a) rigid or TP(b) classification subject to the disposition of the rooflights being in accordance with Table 11 and with the guidance concerning external roof surfaces, Section 15, of The Approved Document.

Light diffusers forming part of a ceiling to rooms and circulation spaces only may be unlimited in their extent if classified TP(a) rigid or limited in area and spacing in accordance with Table 11 and Diagram 24 of the Approved Document, if TP(b).

A **suspended** or **stretched-skin ceiling** to a room only may be a TP(a) flexible classification where each panel should not exceed 5 m^2 and be supported on all sides.

REQUIREMENT B3: INTERNAL FIRE SPREAD (STRUCTURE)

1. The building shall be designed and constructed so that, in the event of fire, its stability will be maintained for a reasonable period.
2. A wall common to two or more buildings shall be designed and constructed so that it adequately resists the spread of fire between those buildings. For the purposes of this sub-paragraph a house in a terrace and a semi detached house are each to be treated as a separate building.
3. To inhibit the spread of fire within the building, it shall be subdivided with fire-resisting construction to an extent appropriate to the size and intended use of the building.
4. The building shall be designed and constructed so that the unseen spread of fire and smoke within concealed spaces in its structure and fabric is inhibited.

The wording of Requirement B3 seeks to apply fundamental fire safety principles to buildings, which can be reiterated as follows:

- provision of fire resistance to elements of structure;
- subdivision/compartmentation of the building;
- maintenance of the integrity of elements at openings; and
- inhibition of fire and smoke spread within concealed spaces.

B3 is split into sections reflecting these principles, with an additional section covering car parks and shopping complexes.

Section 8: Loadbearing elements of structure

The need for a minimum period of fire resistance to the elements of structure of a building is to minimize the risk to:

- occupants remaining in the building,
- people in the vicinity of the building, and
- firefighters occupied in search/rescue operations.

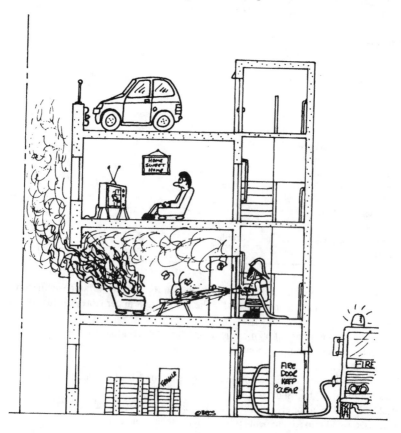

Table 4.9   Minimum periods of fire resistance

Purpose group of building	Minimum periods (min) for elements of structure in a:					
	Basement storey[1] including floor over		Ground or upper storey			
	Depth (m) of a lowest basement		Height (m) of top floor above ground, in building or separating part of building			
	More than 10	Not more than 10	Not more than 5	Not more than 18	Not more than 30	More than 30
1. Residential (domestic):						
(a) flats and maisonettes	90	60	30^2	$60^{3,7}$	90^3	120^3
(b) and (c) dwellinghouses	Not relevant	30^2	30^2	60^8	Not relevant	Not relevant
2. Residential:						
(a) institutional[4]	90	60	30^2	60	90	120^5
(b) other residential	90	60	30^2	60	90	120^5
3. Office:						
not sprinklered	90	60	30^2	60	90	Not permitted
sprinklered	60	60	30^2	30^2	60	120^2
4. Shop and commercial:						
not sprinklered	90	60	60	60	90	Not permitted
sprinklered	60	60	30^2	60	60	120^5
5. Assembly and recreation:						
not sprinklered	90	60	60	60	90	Not permitted
sprinklered	60	60	30^2	60	60	120^2
6. Industrial:						
not sprinklered	120	90	60	90	120	Not permitted
sprinklered	60	60	30^2	60	90	120^5
7. Storage and other non-residential:						
(a) any building or part not described elsewhere:						
not sprinklered	120	90	60	90	120	Not permitted
sprinklered	90	60	30^2	60	90	120^5
(b) car park for light vehicles:						
(i) open-sided park	Not applicable	Not applicable	$15^{2,6}$	$15^{2,6}$	$15^{2,6}$	60
(ii) any other park	90	60	30^2	60	90	120^5

Notes:

[1] The floor over a basement (or if there is more than one basement, the floor over the topmost basement) should meet the provisions for the ground and upper storeys if that period is higher.

[2] Increased to a minimum of 60 min for compartment walls separating buildings.

[3] Reduced to 30 min for any floor within a maisonette, but not if the floor contributes to the support of the building.

[4] Multi-storey hospitals designed in accordance with the NHS Firecode documents should have a minimum 60 min standard.

[5] Reduced to 90 min for elements not forming part of the structural frame.

[6] Increased to 30 min for elements protecting the means of escape.

[7] Refer to Section 7 text.

[8] 30 min in the case of three-storey dwelling houses, increased to 60 min minimum for compartment walls separating buildings.

The first task is to establish the **elements of structure** of the building, namely: structural frames, beams, columns, loadbearing walls (internal and external), floor and gallery structures. **Excluded** from this definition are:

- a structure that only supports a roof (including the roof itself) unless it acts as a floor (e.g. a car park or escape route) or if it is essential for the stability of an external wall requiring fire resistance (e.g. by providing restraint);
- the lowest floor of a building; and
- a platform floor (i.e. a raised or access floor containing services).

Reference can now be made to Table 4.9 where the minimum periods of fire resistance are specified. The table brings together *all* purpose groups and relates period of fire resistance to the storey height above and/or below ground level. Single storey is taken under the heading 'not more than 5'.

As with a number of tables in the Approved Document, careful consideration needs to be given to the supporting notes and the contents of Appendix A. The major items are listed for information:

- Depending on the location of the element of structure it will need to satisfy the fire test criteria with regard to **loadbearing capacity (resistance to collapse), integrity** (resistance to fire penetration) and **insulation** (resistance to heat transfer). Table A1 of the Approved Document verifies which criteria apply to a given part of the building and the method of test exposure. For example, a compartment wall separating different occupancies should offer a minimum 60 min period of fire resistance or as stated by Table 4.9 (whichever is less): the adopted construction will need to satisfy all three test criteria, tested from each side separately. A further example is a cavity barrier, which again should be tested from each side separately but only needs to offer 30 min integrity and 15 min insulation.
- The design and installation of an automatic sprinkler system fitted in accordance with BS 5306: Part 2 can be utilized, for purpose groups 3–7, to give a 30 min reduction in certain cases and are indeed needed where the building exceeds 30 m to the top storey. Note that the system must cover the complete building and take account of the relevant occupancy and the additional requirements for life safety.
- Where one element of structure supports or offers stability to another it shall have at least the period of fire resistance of that other element.
- An element of structure forming part of one or more compartments should adopt the greater relevant period of fire resistance.

- Basement levels open to the external air on at least one side may adopt the period of fire resistance appropriate to the ground or upper storey.
- Certain elements in a single-storey building may not need any fire resistance, although protection may be necessary to external walls (and supports) to limit the extent of unprotected areas or if the element supports a gallery.

In the conversion of two-storey single-family dwellings to provide room(s) in the roof it is often found that the existing first-floor construction only offers a **modified half-hour** fire resistance, 15 min only for integrity and insulation. This is acceptable where only one storey is added, which contains no more than two habitable rooms, does not exceed 50 m² in area, separates rooms only and complies with the means of escape guidance in Section 1.

Raised free standing floors where occupation by persons is limited and does not include members of the public need not have the appropriate period of fire resistance, i.e. to allow an unprotected steel frame. The conditions are that the structure has only one tier and is used for storage purposes only, not more than 10 m in width or length, does not exceed half of the floor area in which it is located, open above and below and means of escape comply with Sections 4, 5 and 6. The introduction of automatic detection to the lower level allows floor sizes up to 20 m in width or length. The floor area would not be limited if an automatic sprinkler system is fitted throughout the building.

The conversion of a house into flats can cause problems with regard to the suitability of the existing timber floors. Therefore for buildings up to three storeys a 30 min period is acceptable; for four storeys and above the full standard should be complied with. Cross reference may also be needed to Approved Document E – Resistance to the passage of sound.

Section 9: Compartmentation

Subdivision into compartments is a long-established principle and takes into account fire load, ease of evacuation and suppression of fire growth. Compartmentation, utilizing walls and floors of the appropriate fire resistance, should be provided where circumstances dictate. These can be summerized in groups as follows:

- a wall common to two or more buildings,
- the separation of different purpose groups,
- the enclosure of places of special fire hazard
- the separation of different occupancies within shop and commercial, industrial or storage premises only,

- the separation of a terrace or semi-detached house,
- the separation of an attached/integral domestic garage,
- any wall or floor separating flats and/or other parts of the building,
- a wall enclosing a refuse storage chamber,
- all floors in other residential and institutional purpose groups,
- any wall needed to limit multi-storey hospital compartments to 2000 m^2 and single-storey hospital compartments to 3000 m^2,
- any wall needed to divide storeys of health care buildings into at least two compartments,
- any wall needed to subdivide a building with reference to Table 4.10,
- any floor where a storey has a floor over 30 m in height,
- floor over basement and floors within basement if more than 10 m deep,
- separation of shopping complex (see BS 5588: Part 10).

For a two storey building, shop, commercial or industrial, the ground storey may be treated as a single storey building subject to:

- upper storey is ancillary and does not exceed 20% of the ground storey area, or 500 m^2, whichever is less, and
- the upper storey is compartmented and provided with independent means of escape routes.

Compartment walls should generally be taken full height in a continuous vertical plane. Also note that only compartment walls and floors in hospitals designed to the Firecodes need to be constructed with materials of limited combustibility where a 60 min or above period of fire resistance applies. Compartment wall or floor junctions need to be carefully considered so as to preserve the integrity of each compartment. A compartment wall, at its junction with a roof, should either continue 375 mm above the roof covering or be fire-stopped to the underside of the covering as detailed in Diagram 28 of the Approved Document, where the roof covering 1.5 m either side of the compartment wall should be designated AA, AB or AC.

The limitation and protection of openings through lines of compartmentation is particularly important so as to restrict the spread of fire. Compartment walls separating buildings or occupancies should be limited to openings for:

- a door for means of escape purposes that has the appropriate fire resistance; and
- the passage of a pipe meeting the provisions of Section 11.

Openings in other compartment walls and floors should be limited to:

- doors that have the appropriate fire resistance;
- the passage of pipes, ventilation ducts, chimneys, appliance

Table 4.10 Maximum dimensions of building or compartment (non-residential buildings)

Purpose group of building or part	Height of floor of top storey above ground level (m)	Floor area of any building or any compartment — In multi-storey buildings	One storey in (m²) — In single storey buildings
Office	No limit	No limit	No limit
Assembly and recreation, shop and commercial:			
(a) schools	No limit	800	800
(b) shops - not sprinklered	No limit	2000	2000
shops sprinklered	No limit	4000	No limit
(c) elsewhere - not sprinklered	No limit	2000	No limit
elsewhere - sprinklered	No limit	4000	No limit
Industrial - not sprinklered	Not more than 18	7000	No limit
	More than 18	2000	No limit
Industrial - sprinklered	Not more than 18	14000	No limit
	More than 18	4000	No limit

		Maximum volume (m³) — In multi-storey buildings	Compartment — In single storey buildings
Storage and other non-residential:			
(a) car park for light vehicles	No limit	No limit	No limit
(b) any other building or part: not sprinklered	Not more than 18	20000	No limit
	More than 18	4000	No limit
sprinklered	Not more than 18	40000	No limit
	More than 18	8000	No limit

Note: Certain industrial and storage uses may be controlled by other legislation.
Crown copyright is reproduced with the permission of the Controller of Her Majesty's Stationery Office.

ventilation ducts or ducts encasing flue pipes meeting the provisions of Section 11;

- refuse chutes of non-combustible construction; and
- atria designed in accordance with BS 5588: Part 7, where atrium breaches compartmentation (reference should also be made to BRE Report 368: *Design methodologies for smoke and heat exhaust ventilation*); and
- protected shafts.

Protected shafts are used to enclose and protect stairs, lifts, escalators, chutes, ducts and pipes passing through lines of compartmentation. The construction and specification of these shafts, which may be horizontal or vertical, should take account of the following items.

- The shaft needs to offer a complete fire barrier to the compartments it serves.
- It should be constructed to offer the appropriate period of fire resistance, noting that this should be achieved from both sides.
- It may include sanitary accommodation or washrooms.
- The incorporation of uninsulated glazed screens.
- It may not contain an oil service pipe (unless it serves a hydraulic lift) or a ventilation duct (unless it pressurizes the stairway) where the shaft also contains stairs or a lift.
- It may contain a natural gas service pipe of screwed steel.
- Adequate ventilation should be provided if the shaft is used to convey flammable gas.
- The external wall element generally need not offer the required period of fire resistance.
- Openings should be protected in line with general compartmentation principles.

Section 10: Concealed spaces (cavities)

The unseen spread of fire and smoke within concealed cavities has resulted in significant damage to buildings and the loss of life. The provisions within this section aim to interrupt cavity firepaths around fire-resisting elements and subdivide extensive cavities by means of cavity barriers. These should offer at least 30 min fire resistance and be adequately specified and installed to take account of cavity dimension and surrounding construction. The first step is to establish if cavity barriers are needed by referring to Table 4.11.

The provisions in Table 4.12 do not apply to a cavity located in one of the following situations:

- in a wall needing fire resistance only because it is loadbearing;
- in a brick, block or concrete cavity wall, with minimum leaf widths

Table 4.11 Cavity barrier provisions

Consider cavity barrier provisions to the following locations:	Applicable purpose group:			
	1a	1b and c	2	2–7
Compartment wall separating buildings	O	O	O	O
Above a protected stairway in a house of three or more storeys	–	O	–	–
External cavity wall junction with all compartment walls and floors	O	–	O	O
Cavity wall junction with all compartment walls and floors and other fire-resisting barriers	O	–	O	O
Above a protected escape route	O	–	O	O
Above bedroom partitions	–	–	O	–
Above a corridor needing to be subdivided for means of escape purposes (please refer to Figure 4.10)	–	–	O	O
Subdivision of cavities to accord with Table 4.12	–	–	O	O
Within rainscreen cladding, to building with a floor above 18 m	O	–	O	–
Edges of cavities and around openings	O	O	O	O

Notes: The symbol O indicates that the provisions apply.
Table 4.11 refers to Table 4.12, which can be checked to verify the maximum dimensions of cavities allowable in non-domestic buildings (purpose groups 2–7).

of 75 mm and maximum cavity width of 100 mm, where the cavity is closed at the top of the wall and at the top of each opening;
- in a floor or roof cavity not exceeding a dimension of 30 m above a **fire-resisting ceiling** extending throughout the building or compartment and constructed to accord with Diagram 31 of the Approved Document;
- below a floor next to the ground where the cavity does not exceed 1 m in height or where it is not normally accessible;
- within an underfloor service void;
- behind rain-screen or over-cladding systems subject to the cavity containing no combustible insulation and the provisions of Table 4.11;
- between double-skinned corrugated or profiled **roof sheeting** being of materials of limited combustibility, where the insulation material used is Class 0 or 1 and makes contact with both inner and outer sheets;
- above any single room exceeding the dimensions of Table 4.12 where the ceiling cavity above has cavity barriers to the room perimeter and at 40 m intervals;

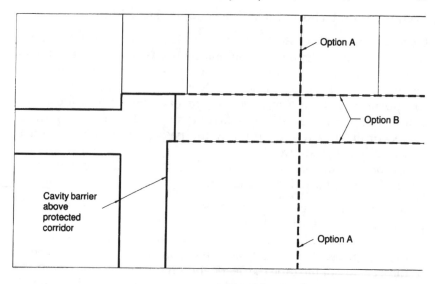

adopt option A or option B above a
corridor which requires sub-division
for means of escape purposes

Figure 4.10 Cavity barrier corridor options.

- above an undivided area that exceeds 40 m in either plan dimension subject to the room and cavity being within their own compartment, an automatic fire detection and alarm system being installed in the building and cavity surfaces strictly limited. Reference should also be made to BS 5588: Part 9 if the cavity is to be used as a plenum.

Where an opening needs to be formed in a cavity barrier this should be limited to one of the following:

- minimum 30 min fire-resisting doors;
- pipes meeting the provisions of Section 11;
- cables or conduits containing cables; and
- openings or ducts (unless fire-resisting) fitted with a suitably installed automatic fire shutter.

Table 4.12 Maximum cavity dimensions (purpose groups 2–7)

Cavity location	Surface designations within cavity (excluding pipes, cables, conduits, etc.)	Maximum dimension in any direction (m)
Between roof and ceiling	Any	20
Any other cavity	Class 0 or 1	20
	Any other Class	10

Section 11: Protection of openings and fire-stopping

Once the lines of fire resistance and allowable openings in them have been established, detailed consideration needs to be given to the fire protection at these locations so as to preserve the integrity of the element and thus avoid fire spread. Note that the test criteria outlined in Appendix A of the Approved Document do not address the issue of smoke spread: for example, a fire damper may offer the required period of fire resistance but allow the passage of smoke, especially in the early stages of fire.

For pipe openings in lines of compartmentation and in cavity barriers three alternatives are given:

- adoption of a proprietary sealing system, e.g. an intumescent collar, for a pipe of any diameter; *or*
- the restriction of the pipe diameter (see Table 15 and Diagram 34 of the Approved Document), e.g. 160 mm for cast iron or steel; *or*
- the sleeving of the pipe (i.e. lead, aluminium, aluminium alloy, PVC or fibre-cement) 1 m either side with a non-combustible material.

For ventilation ducts specific reference should be made to BS 5588: Part 9: 1989 *Code of practice for ventilation and air conditioning ductwork*, which offers a series of alternatives so as to preserve compartment integrity.

For flues, ducts containing flues, and appliance ventilation ducts passing through or forming part of a line of compartmentation, the flue wall should achieve at least half the required period of fire resistance for that wall or floor and be of non-combustible construction. See also Approved Document J – Heat-producing appliances.

Suitable fire-stopping provisions should be provided at:

- joints of fire resisting elements;
- openings around pipes, ducts, conduits and cables; and
- where otherwise specified in the Approved Document.

Fire-stopping needs to take account of thermal movement and unsupported spans exceeding 100 mm. Suitable methods include proprietary sealing systems, intumescent mastics, cement mortar, gypsum-based plaster and glass fibre or crushed rock products.

Section 12: Car parks and shopping complexes

Car parks used for cars and other light vehicles are known to have a fire load that is not particularly high, and fire spread, subject to ventilation arrangements, is likely to be limited. In these circumstances a number of provisions are made, as follows.

- The relevant provisions of requirements B1 and B5 should be applied.

- With the exception of surface finishes, fire doors and attendant's kiosk up to 15 m², and shop mobility facilities, non-combustible construction materials should be used.
- Where the car park does not contain basement levels and is naturally cross-ventilated by permanent openings (5% of the floor area) it may be regarded as **open-sided** with respect to the applicable period of fire resistance (Table 4.9).
- If the car park cannot be regarded as open-sided then natural cross-ventilation (2.5% of the floor area) should be provided which may utilize ceiling level vents; or
- an independent mechanical ventilation system could be utilized, operating at six air changes per hour for petrol vapour extraction and 10 air changes per hour under a fire condition and designed in two parts, each able to extract 50% of the stated rates and each with its own power supply. Extract outlet locations should be equally split at high and low levels, with the fans and ductwork suitably rated for the high temperatures involved. Reference can also be made to BS 7346: Part 2 and BRE Report 368, which offers an alternative approach.
- See also Approved Document F for normal ventilation of car parks.

For individual shops the guidance contained in the Approved Document, and also in BS 5588: Part 11, can be applied. However, where a **shop becomes part of a complex**, which may include covered malls, common service areas and atria for example, then reference should be made to the more detailed guidance found in BS 5588: Part 10: 1991 *Code of practice for enclosed shopping complexes*. This identifies alternative measures and additional compensatory features that are likely to be needed. For information these would include:

- unified management of the complex;
- adequate means of escape;
- smoke control – see BRE Report 386 *Design methodologies for smoke and heat exhaust ventilation* (1999);
- compartmentation;
- sprinkler provisions;
- fire alarm systems;
- access for firefighting;
- isolation of building to relevant boundaries etc.

On a similar principle to restricting fire spread over internal surfaces, the requirement seeks to control heat release rates over external surfaces. The extent of thermal radiation that could pass through a wall is controlled by limiting the extent of unprotected areas. Roof constructions are controlled to restrict fire spread over their surface and penetration from an external fire source.

REQUIREMENT B4: EXTERNAL FIRE SPREAD

1. The external walls of the building shall adequately resist the spread of fire over the walls and from one building to another, having regard to the height, use and position of the building.
2. The roof of the building shall adequately resist the spread of fire over the roof and from one building to another, having regard to the use and position of the building.

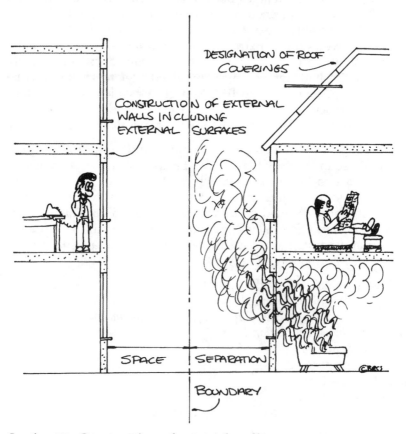

Section 13: Construction of external walls

External walls, as elements of structure, should be able to offer the minimum required period of fire resistance. This is not necessary where the element can be regarded as an **unprotected area**, as defined in Section 14.

Table 4.13   Provisions for external wall surfaces

Dimension to relevant boundary (m)	Classification for building height of:	
	Less than 18 m	*18 m or more*
Less than 1	Class 0	Class 0
1 or more	No provision, except for purpose group 5 buildings exceeding one storey which require an Index, $I < 20$, up to 10 m above ground	Index, $I < 20$, up to 18 m, and Class 0 for wall surfaces over 18 m above the ground

Notes: The index (I) relates to test specifications in BS 476: Part 6. External insulation (and framework) should be of limited combustibility in a building with a storey above 18 m. For further reference see BRE Report 135 *Fire performance of external thermal insulation for walls of multi-storey buildings* (1988).

The external surfaces of walls should now be considered, including the cavity within 'rainscreen' cladding. A relationship is established for any building, relating boundary isolation distance(s) to the building height and allowable external wall surface classifications as indicated in Table 4.13.

An alternative approach could be BRE Fire Note 9 Assessing the fire performance of external cladding systems: a test method (BRE 1999).

Portal frames are widely used in industrial and commercial buildings and can be regarded as acting as single structural elements. In certain circumstances, an external wall close to a boundary for example, these elements of structure may need to offer the required period of fire resistance, which would normally include the rafter section. It has been established that the method of failure under fire loading does not generally warrant the need for fire protection to the rafter section subject to certain provisions, including a satisfactory foundation design. The approved document contains no guidance in this regard but makes reference to the second edition of *The Behaviour of Steel Portal Frames in Boundary Conditions* (1990), available from the Steel Construction Institute. The SCI recommendations need not be followed if the building is fitted with a sprinkler system, incorporating the life safety requirements.

Section 14: Space separation

So as to reduce the risk of fire spread across a boundary from one building to another the guidance limits the extent of external wall openings or **unprotected areas** in relation to the isolation distance of the building to the **relevant boundary** (including a **notional boundary**). The first step with any proposal is to establish the location of the relevant boundaries,

Table 4.14 Method 1 – Small residential buildings, purpose groups 1 and 2

Minimum isolation distance between building and relevant boundary (m)	Maximum unprotected areas (m²)
1	5.6
2	12
3	18
4	24
5	30
6	No limit

Note: Building not to exceed 3 storeys or 24 m in length.

which are those coinciding, parallel to or at an angle not exceeding 80° from the building and may be taken to the centreline of a road, railway, canel or river. The space separation of buildings on the same site is not normally regarded as a risk from a building regulation point of view. An exception is if either of the buildings facing each other, new or existing, are in the residential or assembly and recreation purpose groups: if so, a **notional boundary** needs to be located between the buildings so that both will comply with the provisions for space separation.

Unprotected areas are external wall openings, including windows and doors, external wall panels that do not offer the required period of fire resistance, and external walls with a combustible surface (excluding Class 0) more than 1 mm thick (where half the actual area is regarded as unprotected). Certain unprotected areas can be ignored for calculation purposes:

Table 4.15 Method 2 – Other small buildings or compartments not exceeding 10 m in height

Minimum isolation distance between building and relevant boundary (m)		Maximum unprotected area percentage
Residential, office, assembly and recreation, open-sided car park	Shop and commercial, Industrial, storage and other non-residential	
N/A	1	4
1	2	8
2.5	5	20
5	10	40
7.5	15	60
10	20	80
12.5	25	100

Note: The isolation distances may be halved if the building is fitted throughout with an automatic sprinkler system to BS 5306: Part 2, subject to a minimum distance of 1 m.

- small areas of 0.1 m² and 1 m² subject to dimensional restrictions as outlined in Diagram 44 of the Approved Document;
- the external wall of a protected shaft containing a stairway;
- the external wall of a large uncompartmented building where it is more than 30 m above mean ground level; and
- those in an open-sided canopy, minimum 1 m from the relevant boundary.

The Approved Document contains **Methods 1 and 2** to calculate acceptable limits of unprotected area (Tables 4.14 and 4.15) and refers to two other methods, each offering more refined results, which are contained in BRE Report 187 *External fire spread: building separation and boundary distances* (1991).

Method 3 – Enclosing rectangles (or geometric method) is contained in Part 1 of BRE Report 187. A plane of reference is chosen and the smallest rectangle (taken from tables in the report) is drawn around the extent of unprotected areas to a particular elevation or compartment. The unprotected areas are expressed as a percentage of the rectangle, from which an isolation distance can then be read from the tables. Reduced isolation distance values for residential, office and assembly and recreation, which are a lower fire risk, are quoted in brackets.

Method 4 – Aggregate notional areas (or protractor method) is also contained in Part 1 of BRE Report 187. This method views the building from a series of points along the boundary and calculates the visible unprotected areas by multiplying the notional areas involved by factors that are dependent on boundary isolation distances. A special 'protractor' is used for this purpose. Aggregate notional area limits are given for residential, office and assembly and recreation uses (210 m²) and for shop and commercial, industrial, storage and other non-residential uses (90 m²).

Method 5 – Heat radiation and building separation is the final method referred to in Part 2 of BRE Report 187. The method applies first-principle techniques to establish the potential radiation hazard that one building may have on another and to calculate a satisfactory isolation distance to the relevant boundary or between the buildings. This is the most refined method and can give rise to the most accurate isolation distances.

Section 15: Roof coverings

The guidance within this section aims to restrict fire spread across the roof covering of a building, and the penetration of that covering, where it has been exposed to fire from the **outside**. Provisions are made to limit the use of roof coverings dependent on their **designation** (under test to BS 476: Part 3: 1958, first letter indicating penetration and second letter flame spread) and isolation distance to the **relevant boundary** (see Table 17 of the Approved Document). For example, no restrictions apply to a roof covering designated AA, AB or AC, although if a BA, BB or BC designated covering is to be used then this should be at least 6 m from the relevant boundary. Note that the wall separating a pair of semi-detached houses is not regarded as a relevant boundary for the purposes of Section 15.

For rooflights incorporated within roof coverings consideration should be given to the following aspects.

- Plastic rooflights with at least a Class 3 lower surface or those formed with TP (a) rigid or TP (b) thermoplastic materials should be spaced with at least 3 m between them and accord with the limitations of use stated in Tables 18 and 19 of the Approved Document, where a minimum 6 m dimension to the relevant boundary is required.
- Rooflights formed with rigid thermoplastic sheet products made from polycarbonate or from unplasticized PVC achieving a Class 1 rating for surface flame spread can be regarded as AA designation.
- Rooflights utilizing unwired glass can be regarded as AA designation if the glass is at least 4 mm thick.
- Thatch and wood shingles can be regarded as AD/BD/CD designation. To achieve closer distances to the boundary reference can be made to *Thatched buildings, new properties and extensions* [the "Dorset model"].

REQUIREMENT B5: ACCESS AND FACILITIES FOR THE FIRE SERVICE

1. The building shall be designed and constructed so as to provide reasonable facilities to assist firefighters in the protection of life.
2. Provision shall be made within the site of the building to enable fire appliances to gain access to the building.

Until the introduction of Requirement B5 the fire authority would control the provisions necessary to ensure vehicle and personnel access up to and into the building for rescue and firefighting purposes. The building control authority now takes on this role, as far as it applies to the building design, of checking compliance with the requirement subject to agreed consultation with the fire authority. The degree of access provisions needed to protect the life of firefighting personnel depends mainly on the size of the building. Four sections address the issues controlled under the requirement. Appendix F addresses the issue of the fire behaviour of insulated core panels used for internal structures.

Section 16: Fire mains

The provision of fire mains within the building enables the connection of hoses for water supply to fight fire internally. **Wet** fire mains are permanently charged with water and normally supplied by tanks and pumps located within the building (with an emergency replenishment facility for pumping fire service appliances). **Dry** fire mains are normally empty and supplied by pumping fire service appliances. Wet or dry fire mains should be provided as follows:

- in fire fighting shafts;
- in buildings with a floor level over 60 m above ground or fire service vehicle access level, where a wet rising main is required; and
- in lower buildings where fire mains are provided, they may be wet or dry.

Fire main outlets should be sited within each firefighting lobby between the accommodation and the firefighting shaft. Specific design guidance for fire mains is contained in BS 5306: Part 1.

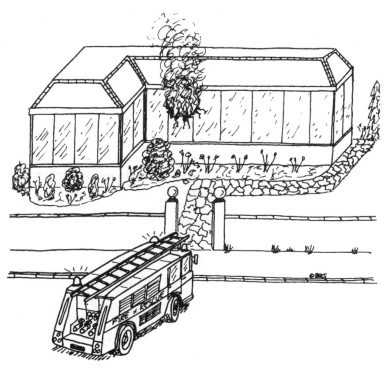

Section 17: Vehicle access

Access to the exterior of the building is required for **pumping appliances** and the deployment of **high-reach appliances** (e.g. turntable
ladders and hydraulic platforms) to floor levels over 11 m above the
ground level, the extent of which depends on building size. For buildings not fitted with fire mains vehicle access should be provided in
accordance with Table 4.16, which relates floor area and height to give
the percentage of the building perimeter that requires access.

For buildings fitted with dry fire mains, access for pumping appliances should be provided to within 18 m of all main inlets, which
should be visible from the appliance. Pumping appliances should
also be able to get within 18 m of a building fitted with wet fire
mains and within sight of the entrance giving to the main and any
inlet for emergency replenishment.

The design of access routes and hardstandings, including any
manholes, etc., should accord with the specifications contained in
Table 4.17.

Table 4.16 Fire appliance access to buildings not fitted with fire mains (excluding
blocks of flats)

Total building floor area (m²)	Height of top storey floor level above ground level (m)	Accessible perimeter percentage required
Up to 2000	Up to 11 (pumping appliance)	See note below
	Over 11 (high-reach appliance)	15
2000–8000	Up to 11 (pumping appliance)	15
	Over 11 (high-reach appliance)	50
8000–16000	Up to and over 11	50
16000–24000	Up to and over 11	75
Over 24000	Up to and over 11	100

Note: These small buildings should have vehicle access to within 45 m of any point on the
building footprint (entrance door for dwellings) or to 15% of the perimeter, whichever is
less onerous. The footprint is the maximum aggregate plan perimeter, at any level, excluding walls common with other buildings. Interior access should be via a minimum 750 mm-
wide door located within the perimeter percentage, which can be used for normal escape
purposes.

Section 18: Access to building for firefighting personnel

In high buildings, where ladder access becomes a problem, and certain
other building forms, further provisions are necessary to allow effective firefighting operations from within the building. This is generally
achieved by the provision of firefighting shafts, which contain fire-

fighting stairs, lobbies and lifts. In the following circumstances fire-fighting shafts should be provided to a building with:

(a) a floor over 18 m above ground or fire service vehicle access level;
(b) a basement over 10 m below ground or fire service vehicle access level;
(c) a storey, exceeding 900 m^2, over 7.5 m above ground or fire service vehicle access level, purpose groups 4, 6 and 7a only;
(d) two or more basement storeys, each exceeding 900 m^2.

Each firefighting shaft should incorporate a firefighting stairway and lift (except items (c) and (d) above) approached via a firefighting lobby containing a fire main outlet (except at access level). For specific design guidance reference should be made to BS 5588: Part 5 (and Part 10 concerning shopping complexes). The guidance is modified for blocks of flats due to the high degree of inherent comparmentation.

The minimum number of firefighting shafts required for a building fitted with sprinklers, to BS 5306: Part 2, is one for a qualifying area up to 900 m^2, two between 900 and 2000 m^2 and for qualifying areas exceeding 2000 m^2, two plus one for every 1500 m^2 or part thereof. For buildings not fitted with sprinklers a firefighting shaft should be provided for each 900 m^2, or part thereof, of the largest qualifying floor area. This criterion also applies to the number of shafts required to serve basement storeys. Their location relates to a 60 m floor isolation, or hose roll-out, distance (40 m where the layout is not known).

Note that rolling shutters in compartment walls should have the ability to be opened and closed by the fire service.

Section 19: Venting of heat and smoke from basements

Fires within basement levels can cause serious problems for fire service personnel where smoke and heat would tend to exit via stairways. Suitably located smoke outlets connecting with the external air should therefore be provided for fire service use to each basement level or compartment with the exception of:

- a basement in a single-family dwelling house, purpose group 1 (b) or 1 (c), or
- any basement not exceeding a floor area of 200 m^2 and 3 m below ground level.

Natural smoke outlets should be located at high level to induce cross-ventilation and offer a total minimum clear cross-sectional area of 2.5% of floor area served. Areas of special fire hazard should have their own independent vents.

A suitably designed **mechanical smoke extract** system can be used as an alternative, offering a minimum ten air changes per hour, subject

Table 4.17 Fire appliance access route specification

Minimum requirements	Pumping appliance	High-reach appliance
Road width between kerbs (m)	3.7	3.7
Gateway width (m)	3.1	3.1
Turning circle between kerbs (m)	16.8	26.0
Turning circle between walls (m)	19.2	29.0
Clearance height (m)	3.7	4.0
Carrying capacity (t)	12.5	17.0

Note: Any 'dead-end' access route exceeding 20 m should be provided with a turning circle or hammerhead to accord with the above criteria. For high-reach appliances a clear overhead zone should be maintained in front of the building elevations which require access, maximum dimension 10 m deep (see Diagram 49 of the Approved Document).

to the basement levels also being provided with a sprinkler system to accord with BS 5306: Part 2.

Basement car parks can use the guidance in Section 12.

It should be noted that provisions are not made in the Approved Document for smoke venting to ground and upper floor levels. Certain local Acts and British Standards may require such provisions, which normally seek a clear ventilation area of 2.5% of the floor area in question and the clearance of smoke from stairways. Atria, shopping complexes and buildings to which a fire safety engineering approach have been applied may also require the provision of some form of natural or mechanical smoke control measures.

APPENDICES

The principal contents of the appendices contained within the Approved Document have been touched upon within the body of the text; however, the appendix titles and the following specific items have been highlighted for information.

Appendix A: Performance of materials and structures

Appendix B: Fire doors

Appendix C: Methods of measurement

- The measurement of cubic capacity would include any roof space over a compartment.
- The height of the top storey of a building is measured from the upper surface of the top floor (excluding rooftop plant areas) down to ground level on the lowest side of the building.

- The height of a building is measured from the mean roof level down to the mean ground level.

Appendix D: Purpose groups

Table D1 is reproduced earlier within the text as Table 4.1.

Appendix E: Definitions

Most of the definitions concerning Requirements B1–B5 have been highlighted and explained within the text. The following are provided for information:

Atrium – a space within a building, not necessarily vertically aligned, passing through one or more structural floors (excluding a shaft used only for stairs, escalators, lifts or services) also refer to BS 5588: Part 7.

Basement storey – a storey with a floor that at some point is more than 1.2 m below the highest level of ground adjacent to the outside walls.

Circulation space – a space (including a protected stairway) mainly used as a means of access between a room and an exit from the building or compartment.

Common stair – an escape stair serving more than one flat or maisonette.

Compartment (fire) – a building or part of a building, comprising one or more rooms, spaces or storeys, constructed to prevent the spread of fire to or from another part of the same building, or an adjoining building.

Direct distance – the shortest distance from any point within the floor area, measured within the external enclosures of the building, to the nearest storey exit ignoring walls, partitions and fittings, other than the enclosing walls/partitions to protected stairways.

Dwelling – a unit of residential accommodation occupied (whether or not as a sole or main residence):
(a) by a single person or by people living together as a family; or
(b) by not more than six residents living together as a single household, including a household where care is provided for residents.

Emergency lighting – lighting provided for use when the supply to the normal lighting fails.

Escape lighting – the part of emergency lighting that is provided to ensure that the escape route is illuminated at all material times.

Escape route – forming part of the means of escape from any point in a building to a final exit.

External wall (or **side of a building**) – includes a part of a roof pitched at an angle of more than 70° to the horizontal, if that part of the roof adjoins a space within the building to which persons have access (but not access only for repair or maintenance).

Fire door – a door or shutter (including a cover or other form of protection to an opening), provided for the passage of persons, air or objects, which together with its frame and furniture as installed in a building, is installed (when closed) to resist the passage of fire and/or gaseous products of combustion, and is capable of meeting specified performance criteria to those ends.

Fire-resisting (fire resistance) – the ability of a component or construction of a building to satisfy, for a stated period of time, some or all of the appropriate criteria specified in the relevant part of BS 476.

Fire separating element – a compartment wall, compartment floor, cavity barrier and construction enclosing a protected escape route and/or a place of special fire hazard.

Gallery – a floor (including a raised storage area) that is less than one-half of the area of the space into which it projects.

Material of limited combustibility – a material performance specification that includes non-combustible materials, and for which the relevant test criteria are set out in Appendix A, paragraph 8 (and Table A7) of the Approved Document.

Means of escape – structural means whereby (in the event of fire) a safe route or routes is or are provided for persons to travel from any point in a building to a place of safety.

Non-combustible material – the highest level of reaction to fire performance. The relevant test criteria are set out in Appendix A, paragraph 7 (and Table A6) of the Approved Document.

Open spatial planning – the internal arrangement of a building in which more than one storey or level is contained in one undivided volume, e.g. split level floors. For the purposes of this document there is a distinction between open spatial planning and an atrium space.

Platform floor – a floor (including an access or raised floor) supported by a structural floor, but with an intervening concealed space that is intended to house services.

Protected corridor/lobby – a corridor or lobby that is adequately protected from fire in adjoining accommodation by fire-resisting construction.

Protected entrance hall/landing – a circulation area consisting of a hall or space in a dwelling, enclosed with fire-resisting construction (other than any part that is an external wall of a building).

Protected stairway – a stair discharging through a final exit to a place of safety (including any exit passageway between the foot of the stair and the final exit) that is adequately enclosed with fire-resisting construction.

Storey – includes:
  (a) any gallery in an assembly building, purpose group 5; and
  (b) any gallery in any other type of building if its area is more than half that of the space into which it projects; and
  (c) a roof, unless it is accessible only for maintenance and repair.

Storey exit – a final exit, or a doorway giving direct access into a protected stairway, firefighting lobby, or external escape route.

Travel distance – the actual distance to be travelled by a person from any point within the floor area to the nearest storey exit, having regard to the layout of walls, partitions and fittings (unless otherwise specified, e.g. as in the case of flats).

Appendix F: Fire behaviour of insulating core panels used for internal structures

The introduction of this appendix reflects the recent and extensive research undertaken on the subject. Insulating core panels are now widely used for a range of internal structures. Examples include cold rooms and 'clean' environment enclosures. The panels forming these free standing/structural enclosures are generally of sandwich construction with facing sheets of galvanized steel. Insulants used include, expanded or extruded polystyrene, polyurethane, mineral fibre, polyisocyanurate and modified phenolic.

Once exposed to heat from a fire, polymeric core materials (not mineral fibre) generate large quantities of smoke. The overall integrity of the system used under fire conditions requires careful consideration. General design guidance is given within Appendix F, where a risk assessment approach is recommended. Specific reference is also made to *Design, construction, specification and fire management of insulated envelopes for temperature controlled environments* published by the International Association of Cold Storage Contractors (European Division).

Examples of core materials for a range of applications is given:

- Mineral fibre – cooking areas, hot areas, bakeries, fire breaks, fire stop panels and general fire protection.
- All cores – chill and cold stores, blast freezers, food factories and clean rooms.

Appendix G: Standards and other publications referred to

The majority of publications relevant to Requirements B1–B5 have been listed within the text. The following titles are now included for consistency:

- Joint circulars DOE 12/92 and 12/93/Welsh Office 25/92 and 55/93 *Houses in multiple occupation*, (HMSO).
- Gas Safety (Installation & Use) Regulations 1998.
- Pipelines Safety Regulations 1996, S1 1996 No 825.
- BRE Report 128 *Guidelines for the construction of structural elements* (1988).
- BRE Digest 208 *Increasing the fire resistance of existing timber floors* (1988).
- LPC (1996) *Design guide for the fire protection of buildings.*
- ASFPCM (1992) *Fire protection for structural steel in buildings* (2nd edn).

CHAPTER 5

Approved Document C: Site preparation and resistance to moisture

Requirements C1, C2 and C3 seek to address the risks associated with vegetable matter on the building site, contaminants on or in the ground and groundwater, but only as far as is necessary to ensure the health and safety of persons in buildings. Requirement C4 deals specifically with the resistance of the building to moisture penetration via the floor, walls and roof.

REQUIREMENT C1: PREPARATION OF SITE

The ground to be covered by the building shall be reasonably free from vegetable matter.

REQUIREMENT C3: SUBSOIL DRAINAGE

Adequate subsoil drainage shall be provided if it is needed to avoid:
(a) the passage of ground moisture to the interior of the building;
(b) damage to the fabric of the building.

Section 1: Site preparation and site drainage

Prior to the construction of any building on a site consideration needs to be given to the following fundamental aspects.

Organic material

- Sufficient turf and other vegetable matter should be removed from the site of building, normally to a depth of at least 150 mm. This does not apply in the case of general storage buildings, or a building where the health or safety of those employed would not be affected.
- Building services should be of such design as to protect them against damage from tree roots. For example, below-ground drainage could be suitably encased in concrete, or utilize pipes with flexible joints.

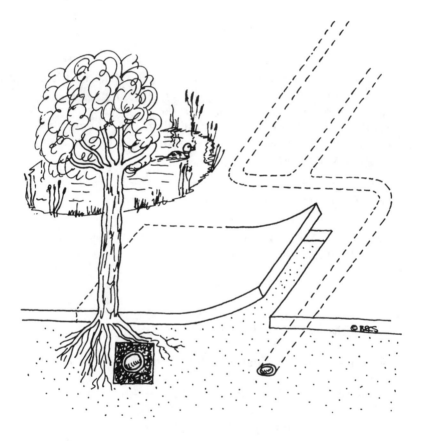

Site drainage

- Where the site is liable to flooding appropriate steps should be taken.
- Effective drainage should be provided to protect the building against a high water table (rising to within 0.25 m of the lowest floor level) and the entry of surface water.

- Where an active subsoil drain is cut, it should be sealed if passing under a building, re-routed around the building or re-run to another outfall.
- An **alternative approach** would be to design the building so as to prevent the passage of ground/surface water; please refer to Sections 3-5 later within the text.

REQUIREMENT C2: DANGEROUS AND OFFENSIVE SUBSTANCES

Reasonable precautions shall be taken to avoid danger to health and safety caused by substances found on or in the ground to be covered by the building.

Section 2: Contaminants

Building sites may contain a wide range of contaminants as a consequence of the previous use(s) of the site, these are illustrated in Figure 5.1.

Contaminants may emanate from one of the following sources:

- solid, liquid and gases arising from previous use of land;
- natural contamination by radon and its decay products; or
- landfill gases from buried waste.

The first stage is to identify land that may contain contaminants. This may be possible from local planning records, the refuse authority or local knowledge: for example, a local authority building control officer may remember the use of the site in the past. Where no records exist, and contaminants in the ground are suspected, reference can be made to Table 5.1, which gives an indication of the signs to look for and actions to be taken.

The relevant actions outlined above take one of three forms:

- **Removal** – remove the contaminant (and any contaminated ground) for a depth of 1 m (or less if agreed with controlling authority).
- **Filling** – cover the contaminant with suitable material to a depth as above.
- **Sealing** – provide a suitably sealed imperforated barrier.

As an **alternative approach** to the above, and in the most hazardous of cases, expert advice should be sought and reference made to BS (DD)

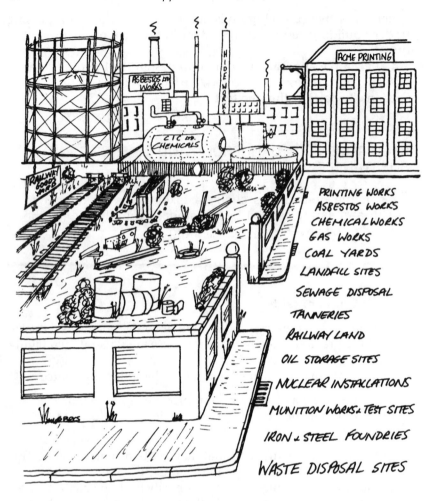

PRINTING WORKS
ASBESTOS WORKS
CHEMICAL WORKS
GAS WORKS
COAL YARDS
LANDFILL SITES
SEWAGE DISPOSAL
TANNERIES
RAILWAY LAND
OIL STORAGE SITES
NUCLEAR INSTALLATIONS
MUNITION WORKS & TEST SITES
IRON & STEEL FOUNDRIES
WASTE DISPOSAL SITES

Figure 5.1 Sites likely to contain contamination.

175: 1988 *Code of practice for the identification of potentially contaminated land and its investigations* and BS 5930: 1981 *Code of practice for site investigations.*

Note that where there are signs of possible contaminants the environmental health officer of the local authority must be notified immediately. The provisions in Section 2 go on to address the problems that need to be overcome with respect to the gaseous contaminants.

Radon, which is a colourless and odourless gas, can migrate through subsoil and in turn into buildings, and excessive exposure can increase the risk of lung cancer. Precautions may therefore be necessary in the following regions:

Table 5.1 Possible contaminants and actions

Signs of possible contaminant	Possible contaminant	Relevant action
Vegetation (absence, poor or unnatural growth)	Metals	None
	Metal compounds*	None
	Organic compounds	Removal
	Gases	Removal
Surface materials (unusual colours and contours may indicate wastes and residues)	Metals	None
	Metal compounds*	None
	Oily and tarry wastes	Removal, filling or sealing
	Asbestos (loose)	Filling or sealing
	Other mineral fibres	None
	Organic compounds (including phenols)	Removal or filling
	Combustible material (including coal and coke dust)	Removal or filling
	Refuse and waste	Total removal or see guidance
Fumes and odours (may indicate organic chemicals at very low concentrations)	Flammable explosive and asphyxiating gases (including methane and CO_2)	Removal
	Corrosive liquids	Removal, filling or sealing
	Faecal animal and vegetable matter (biologically active)	Removal or filling
Drums and containers (full or empty)	Various	Removal with all contaminated ground

Notes: Liquids and gaseous contaminants are mobile, and the ground covered by the building can be affected by such contaminants from elsewhere. Some guidance is given for landfill gas and radon; other liquids and gases should be referred to a specialist. Crown copyright is reproduced with the permission of the Controller of Her Majesty's Stationery Office.

* indicates that special cement may be needed with sulphates.

- Cornwall or Devon;
- parts of Somerset;
- Northamptonshire or Derbyshire.

Current information on specific sites can be obtained from the controlling authority. Construction guidance is contained within the BRE Report *Radon: guidance on protective measures for new dwellings* (1999).

Landfill gas and **methane**, which includes carbon dioxide as a toxic gas, can migrate under pressure through the subsoil and enter the building via cracks. Note that gas from peat and other natural gases are treated in the same way. A general approach can be summarized as follows.

- Where the building will be on or within 250 m of landfill an investigation should be made.
- For methane levels not exceeding 1% by volume a suspended and ventilated concrete floor should suffice (for dwellings only).
- For carbon dioxide levels exceeding 1.5% by volume preventive gas ingress measures are likely to be required. Specific design measures will certainly be necessary over a 5% level (for dwellings only).
- In all other cases, including non-domestic applications, expert advice should be sought. This should include a complete investigation and/or monitoring of the site.

Specific reference should also be made to the BRE Report *Construction of new buildings on gas contaminated land* (1991). Other reference sources for the overall subject are confirmed for information:

- DOE (HM Inspectorate of Pollution): *The control of landfill gas*, Waste Management Paper No. 27, 1989, HMSO.
- BRE Report Measurement of gas emissions from contaminated land (1987), HMSO.
- ICRCL 17/78 *Notes on the development and after-use of landfill sites* 8th edn (1990) (available from DOE Publication Sales Unit).
- ICRCL 59/83 *Guidance on the assessment and redevelopment of contaminated land* 2nd edn (1987) (available from DOE Publication Sales Unit).
- Institute of Wastes Management (1989) *Monitoring of Landfill Gas.*

REQUIREMENT C4: RESISTANCE TO WEATHER AND GROUND MOISTURE

The walls, floors and roof of the building shall adequately resist the passage of moisture to the inside of the building.

Section 3: Floors next to the ground

This section deals with the need for a floor next to the ground to:

- resist the entry of moisture to the upper surface of the floor, but not in the case of general storage buildings or a building where the health or safety of those employed would not be affected; and
- not to be damaged as a consequence of moisture from the ground.

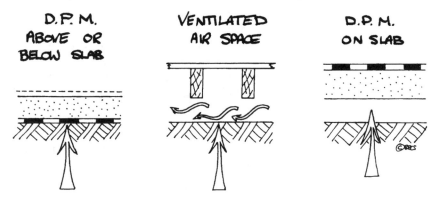

Figure 5.2 Provisions for floors next to the ground.

Three types of floor are dealt with, as illustrated in Figure 5.2.

Floors supported directly on the ground should be constructed to the following guidelines, unless they are subject to water pressure, where the **alternative approach** should be adopted:

- bed of clean hardcore, e.g. broken bricks or reject stone, containing no harmful sulphates, and
- minimum 100 mm thick concrete, BS 5328 mix ST2 (or ST4 if reinforcement is being used) or equivalent, and
- damp-proof membrane (DPM) minimum 300 m (1200 gauge) polythene, or 250 m (1000 gauge) if Agrément Certificate or PIFA standard applies, with joints sealed, lapped with damp-proof courses (DPC) and laid under the slab on a suitable bedding material, or
- the damp-proof membrane laid over the slab, which may be as above or use a three-coat cold-applied bitumen solution protected by the screed or floor finish.
- A timber floor finish directly on a concrete slab should have wood preservative treatment to any timber fixing fillets and be bedded in a material acting as a damp-proof membrane.
- For an **alternative approach** reference can be made to CP 102: 1973 Protection of buildings against water from the ground, or BS 8102: 1990 Code of practice for protection of structures against water from the ground. Reference can also be made to Approved Document – *Basements for dwellings*.

For a suspended timber floor the following guidelines should be adopted.

- Cover the ground with 100 mm concrete on a hardcore base or with a DPM, as described above, weighted with 50 mm concrete or fine aggregate.

- The ground level under the floor should be above the adjoining ground level or effectively drained.
- The ventilated air space, 1500 mm^2 per metre run on both sides, should be at least 150 mm to the underside of the floor from the ground cover level and 75 mm to the underside of any wall plates (supporting floor joists) at which location a DPC should be installed.
- For an **alternative approach** reference can be made to CP 102: 1973 *Protection of buildings against water from the ground.*

Suspended concrete floors should follow these design guidelines and any reinforcement should be protected from moisture:

- Use in-situ concrete, minimum 100 mm thick, 40 mm reinforcement cover; or
- precast concrete, with or without infill blocks, reinforcement protected to at least moderate exposure; and
- incorporate a DPM if ground cover level is below adjoining ground level and not effectively drained; and
- if gas accumulation could give rise to an explosion provide a ventilated air space, 1500 mm^2 per metre run on both sides.

Section 4: Walls

Walls should:

- resist the entry of moisture into the building from the ground and from rain and snow (external walls), but not apply in the case of general storage buildings or a building where the health or safety of those employed would not be affected; and
- not be damaged by such entry of ground moisture and rain and snow (external walls).

Three situations are highlighted in the Approved Document guidance, which is illustrated in Figure 5.3.

For internal and external walls that need to resist moisture from the ground a suitable DPC of bituminous material, engineering bricks or slates should be installed, at least 150 mm above adjoining ground level for an external wall, and lapped with any DPM. For a cavity external wall a cavity tray discharging to the outside face may be required, e.g. over a concrete window beam or to the perimeter of a raft type foundation.

An **alternative approach**, where groundwater pressure may present a problem, can be found in BS 8215: 1991 *Code of practice for design and installation of damp-proof courses in masonry construction* or BS 8102: 1990 *Code of practice for protection of structures against water from the ground.*

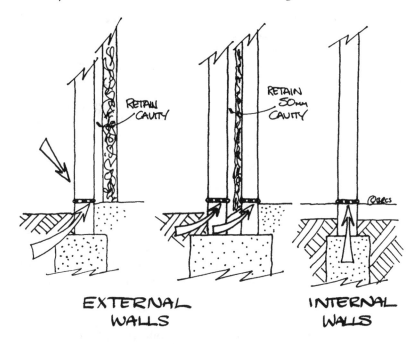

Figure 5.3   Provisions for walls.

Reference can also be made to Approved Document – *Basements for dwellings.*

External walls also need to offer protection from rain and snow: to achieve this one of the following design options may be chosen. For solid walls subject to very severe exposure external cladding should be used, but for those subject to severe exposure or less then adopt the following:

- minimum 328 mm brickwork, 250 mm dense aggregate concrete blockwork or 215 mm lightweight aggregate concrete blockwork; and
- minimum 20 mm two-coat rendering, reference BS 5262: 1976 *Code of practice. External render finishes*; and
- top of the wall to be protected by a coping and a DPC, where necessary.
- The **alternative approach** is to follow BS 5628: Part 3: 1985 for masonry and BS 5390: 1976, for stone masonry.

For cavity external walls:

- the outer leaf should be of brick, block, stone or cast stone; and
- have at least a 50 mm cavity, maintaining a nominal 50 mm residual

cavity where partial insulation fill is used, bridged only by wall ties, cavity tray or vertical DPC at openings; and

- an inner leaf of masonry or lining frame, e.g. timber or steel frame.
- The **alternative approach** is to follow BS 5628: Part 3: 1985, for masonry.

An external wall cavity may be insulated during the course of construction or after the wall is complete. This should be carried out in strict accordance with an Agrément Certificate or equivalent approved system. The suitability of an existing cavity wall should be assessed in accordance with BS 8208: Part 1: 1985.

Section 5: Cladding for external walls and roofs

Cladding can take a number of forms, e.g. steel sheeting to a wall or clay tiling to a roof. Both should resist the penetration of rain and snow to the inside of the building and not be damaged by such moisture.

The four cladding design options can be summarized as follows:

- **impervious**, including metal, plastic, glass and bituminous products; or
- **weather-resisting**, including natural stone or slate, cement-based products, fired clay and wood; or
- **moisture-resisting**, including bituminous and plastic products with joints lapped, if used as a sheet material, and permeable to water vapour unless a ventilated space is provided directly behind the material; or
- **jointless materials** and **sealed joints**, allowing for structural and thermal movement.

Note that the suitability of dry joints used in cladding systems will depend on design and wind/rain exposure. In addition, paint and other surface coatings are not regarded as a weather-resisting part of a wall.

The incorporation of insulation should take into account the need for moisture protection and the problems associated with condensation and cold bridging, which are highlighted in Approved Document F and the BRE Report *Thermal insulation: avoiding risks* (1994).

As an **alternative approach** a number of British Standards are referred to:

- BS CP 143: Code of practice for sheet roof and wall coverings:
- Part 1: 1958 for corrugated and troughed aluminium;
- Part 5: 1964, zinc; Part 10: 1973, galvanized corrugated steel;
- Part 12: 1970 (1988), copper; Part 15: 1973 (1986), aluminium;
- Part 16: 1974, semi-rigid asbestos bitumen sheets.
- BS 6915: 1988 for lead.

- BS 5247: Part 14: 1975 for corrugated asbestos-cement.
- BS CP 297: 1972 Precast concrete cladding (non-loadbearing).
- BS 8298: 1989 for natural stone (1994).
- BS 8200: 1985 for walls and steep roofs in general.

CHAPTER 6

Approved Document D: Toxic substances

This approved document contains just one requirement which deals with the health risk of cavity wall insulating materials that give off formaldehyde fumes.

REQUIREMENT D1: CAVITY INSULATION

If insulating material is inserted into a cavity in a cavity wall reasonable precautions shall be taken to prevent the subsequent permeation of any toxic fumes from that material into any part of the building occupied by people.

The technical solution offered in the Approved Document seeks to minimize, as far as is practicable, the passage of fumes to the occupied parts of the building. A cavity wall may be insulated with urea formaldehyde foam subject to the following provisions.

- The inner leaf should be built of bricks or blocks.
- The cavity wall should be assessed, before filling, in accordance with BS 8208: Guide to assessment of suitability of external cavity walls for filling with thermal insulants Part 1: 1985 Existing traditional cavity construction.
- A current BSI Certificate of Registration of Assessed Capability should be held by the person undertaking the work.
- The material used should accord with BS 5617: 1985 Specification for urea-formaldehyde (UF) foam systems suitable for thermal insulation of cavity walls with masonry or concrete inner and outer leaves.
- Installation should accord with BS 5618: 1985 Code of practice for thermal insulation of cavity walls (with masonry or concrete inner and outer leaves) by filling with urea-formaldehyde (UF) foam systems.

Note that careful consideration should be given to the detailing of the cavity wall construction if the foam is to be installed in recently completed construction. As a combustible insulant reference should also be made to the guidance for Approved Document B, relating to the need for cavity barriers as outlined in Table 4.11.

CHAPTER 7

Approved Document E: Resistance to the passage of sound

The nuisance of sound is regarded as a health and safety issue for persons living in dwellings where the enclosing construction of a dwelling needs to be able to control noise levels emanating from adjoining dwellings, other buildings or other parts of the same building. The occupants of the dwelling should be allowed to follow normal domestic activities, including sleep and rest, without threat to their health from sound sources.

Sound emanates from one of two sources. An **airborne** source, speech or loadspeakers for example, vibrates the surrounding air, which in turn sets up vibrations in the enclosing walls and floors. An **impact** source, including footsteps or the movement of furniture, sets up vibrations by direct contact.

Vibrations from both these sources spread via floors, internal walls and the inner leaves of external walls, vibrating the air next to them, which causes new airborne vibrations, which are then heard as sound. Requirements E1, E2 and E3 require walls, floors and stairs to offer airborne and/or impact sound resistance.

REQUIREMENT E1: AIRBORNE SOUND (WALLS)

A wall which:

(a) separates a dwelling from another building or from another dwelling, or

(b) separates a habitable room or kitchen within a dwelling from another part of the same building which is not used exclusively as part of the dwelling,

shall have reasonable resistance to the transmission of airborne sound.

REQUIREMENT E2: AIRBORNE SOUND (FLOORS AND STAIRS)

A floor or a stair which separates a dwelling from another dwelling, or from another part of the same building which is not used exclusively as part of the dwelling, shall have reasonable resistance to the transmission of airborne sound.

REQUIREMENT E3: IMPACT SOUND (FLOORS AND STAIRS)

A floor or a stair above a dwelling which separates it from another dwelling, or from another part of the same building which is not used exclusively as part of the dwelling, shall have resonable resistance to the transmission of impact sound.

To clarify the application of the provisions of Requirements E1–E3 refer to Figure 7.1.

To insulate the occupants of a dwelling from sound produced from one or both of the sources previously highlighted the flow of sound energy by direct or indirect (flanking) transmission should be restricted. Figure 7.2 illustrates the principles.

To reduce the **direct** and **flanking transmission** of airborne and impact sound, through a wall or floor from one side to another, the Approved Document gives guidance on acceptable methods of construction. Sections 1–4 deal with new build and Sections 5 and 6 cover conversion work in relation to a material change of use. Different forms and materials used in the construction of walls, floors and stairs offer different methods to resist sound. For example, a solid masonry separating wall relies on its mass, which is not readily set into vibration, whereas timber-framed construction can offer lightweight structural isolation.

With respect to conversion work, the guidance offered in the Approved Document recognizes the fact that it is often difficult to improve flanking sound transmission properties in existing construction. A further point of consideration is the advantages obtained when steps or staggers between adjoining dwellings can be incorporated, and varying room layouts.

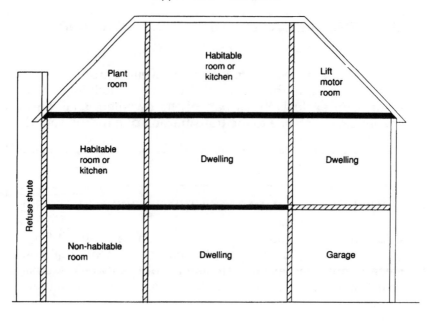

Figure 7.1 Walls and floors required to offer sound resistance.
Note: Regulation 6 applies Requirements E1–E3 to a material change of use where the building is to be used as a dwelling or flat where previously it was not. This also includes a dwelling adjacent to or within another building type.

Sound from external sources is not controlled under Requirements E1–E3. Certain sound levels above those encountered for normal domestic activities may arise: for example, a dwelling located within or adjacent to a night club. In these situations further specialist provisions may be necessary.

Section 1: Separating walls for new building

Walls need only to resist the passage of airborne sound, and four main options to achieve this are given (Figure 7.3). Included with each wall type are example constructions, which can offer the required level of sound insulation, junction details and a checklist of items to consider.

Wall type 1: Solid masonry

Constructions:

- Bricks, laid in a bond including headers, 13 mm plaster on both room faces, mass including plaster 375 kg/m².

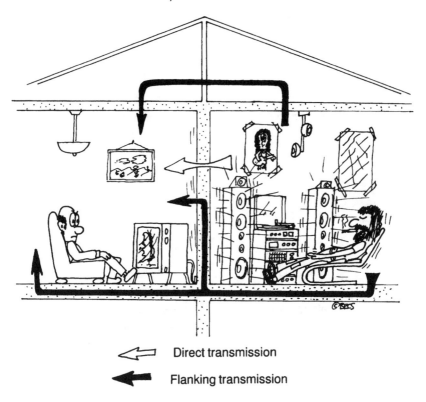

⟨▭ Direct transmission

◀━━ Flanking transmission

Figure 7.2 Sound transmission paths.

- Concrete blocks, to the full thickness of the wall, 13 mm plaster on both room faces, mass including plaster 415 kg/m^2.
- Bricks, laid in a bond including headers, 12.5 mm plasterboard on both room faces, mass including plasterboard 375 kg/m^2.
- Concrete blocks, to the full thickness of the wall, 12.5 mm plasterboard on both room faces, mass including plasterboard 415 kg/m^2.
- Concrete, in-situ or large panel (*minimum* density 1500 kg/m^3), panel joints mortar filled, mass including plaster, if used, 415 kg/m^2.

Junctions:

- Fill (or fire-stop) joint between roof and head of wall.
- Mass of wall within a roof space above a heavy ceiling (12.5 mm plasterboard) may be reduced to 150 kg/m^2. For lightweight aggregate blocks (density less than 1200 kg/m^3) seal one face with cement paint or plaster skim.
- Floor joists may be supported by hangers or built in, subject to good workmanship to avoid air paths. For concrete floor types 1 and 2 the wall or floor may be carried through.

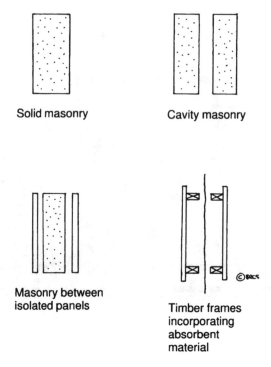

Figure 7.3   Wall types.

- External cavity wall outer leaf may be of any construction.
- Bond or tie the separating wall to an external cavity wall inner leaf of masonry, or a solid external masonry wall, minimum mass 120 kg/m^2 unless openings are provided each side of the separating wall, minimum 1 m high and not more than 700 mm from both faces of the separating wall.
- Tie the separating wall to an external cavity wall inner leaf of timber, joints sealed with tape or caulking.
- Cavity at separating wall location to be stopped, e.g. mineral wool.

Consider:

- Seal/fill all masonry joints with mortar, bricks laid frog up.
- No restrictions apply for partition walls meeting separating wall.

Wall type 2: Cavity masonry

Constructions:

- Two leaves of brick with 50 mm cavity, 13 mm plaster on both room faces, mass including plaster 415 kg/m^2.

- Two leaves of concrete block with 50 mm cavity, 13 mm plaster on both room faces, mass including plaster 415 kg/m^2.
- Two leaves of lightweight aggregate block (*maximum* density 1600 kg/m^3) with 75 mm cavity 13 mm plaster or 12.5 mm plasterboard on both room faces, mass including finish 300 kg/m^2.

Where a minimum 300 mm step or stagger is employed between dwellings the following revised constructions may be used:

- Two leaves of concrete block with 50 mm cavity, 12.5 mm plasterboard on both room faces, mass of masonry *only* 415 kg/m^2.
- Two leaves of lightweight aggregate block (*maximum* density 1600 kg/m^3) with 75 mm cavity 13 mm plaster or 12.5 mm plasterboard on both room faces, mass including finish 250 kg/m^2.

Junctions:

- Fill (or fire-stop) joint between roof and head of wall.
- Mass of wall within a roof space above a heavy ceiling (12.5 mm plasterboard) may be reduced to 150 kg/m^2. For lightweight aggregate blocks (density less than 1200 kg/m^3) seal one face with cement paint or plaster skim.
- Floor joists may be supported by hangers or built in, subject to good workmanship to avoid air paths. Concrete floor should be carried through to the cavity face of each leaf.
- External cavity wall outer leaf may be of any construction.
- Bond or tie the separating wall leaves to an external cavity wall inner leaf of masonry, minimum mass 120 kg/m^2 unless concrete blocks are used for the separating wall.
- Tie the separating wall to an external cavity wall inner leaf of timber, joints sealed with tape or caulking.
- Cavity at separating wall location to be stopped, e.g. mineral wool.

Consider:

- Seal/fill all masonry joints with mortar, bricks laid frog up.
- No restrictions for partition walls meeting separating wall.
- Maintain cavity for the full height of the separating wall, up to the underside of the roof.
- Leaves should only be connected by pattern **butterfly** ties.
- Unbonded external wall insulation material, e.g. polystyrene beads, should be prevented from entering the cavity by a flexible closer, e.g. mineral wool.

Wall type 3: Masonry between isolated panels

- Constructions comprise masonry cores with either of the two panel constructions isolated by a 25 mm air space each side:

- Brick, mass 300 kg/m^2.
- Concrete block, mass 300 kg/m^2.
- Lightweight concrete block (*maximum* density 1600 kg/m^3), mass 160 kg/m^2.
- Cavity brickwork or blockwork, *any* mass.

with:

- two sheets of plasterboard joined by cellular core, mass 18 kg/m^2, fixed to ceiling and floor only, joints taped between panels; or
- two sheets of 12.5 mm plasterboard, joints staggered, onto supporting framework or 30 mm plasterboard where no framework employed.

Junctions:

- Fill (or fire-stop) joint between roof and head of wall.
- Mass of wall within the roof space above a heavy ceiling (12.5 mm plasterboard) may be reduced to 150 kg/m^2 and the panels omitted. For open-textured lightweight aggregate blocks (density less than 1200 kg/m^3) seal one face with cement paint or plaster skim. Seal junction between panels and ceiling with tape or caulking.
- Floor joists to be supported by hangers, and gaps between joists sealed with blocking at line of panels. Concrete floor may be carried through if it has a mass of at least 365 kg/m^2 or stopped at the cavity face of each leaf for a cavity core.
- External cavity wall outer leaf may be of any construction.
- External cavity wall inner leaf may be of any construction if it is lined with isolated panels.
- Cavity at separating wall location to be stopped, e.g. mineral wool.

Consider:

- Seal/fill all masonry joints with mortar, bricks laid frog up.
- Panels should not be fixed or tied to the masonry core construction.
- Partition walls meeting a solid core separating wall should not be of masonry construction. Other loadbearing partitions should be fixed via a continuous pad of mineral wool. Non-loadbearing partitions should be butted tightly against the isolation panels. All joints sealed with tape or caulking.
- Maintain cavity for the full height of the separating wall, up to the underside of the roof, where applicable. Leaves should only be connected by pattern **butterfly** ties.

Wall type 4: Timber frames incorporating absorbent material

- Constructions comprise timber frames, with or without a masonry core:

- Timber frames, 200 mm between linings of 30 mm plasterboard each side, joints staggered with absorbent material suspended in cavity of 25 mm unfaced mineral fibre batts or quilt, 50 mm if fixed to a frame.
- Any masonry core with timber frames, 200 mm between linings of 30 mm plasterboard each side, joints staggered with absorbent material suspended in cavity of 25 mm unfaced mineral fibre batts or quilt, 50 mm if fixed to one of the frames. Core should only connect with *one* frame.

Junctions:

- Fill (or fire-stop) joint between roof and head of wall.
- Carry full wall construction to the underside of the roof and fill (or fire-stop) joint between roof and head of wall. Alternatively carry one frame up only, with minimum 25 mm plasterboard lining to each side, use a non-rigid cavity closer at ceiling level and also seal the space between the frame and roof finish.
- At intermediate floor joist level block air paths by carrying linings through or seal with solid timber blocking at line of panels.
- For an external cavity wall seal the end of the separating wall up to the outer leaf to prevent sound paths.

Consider:

- Timber frames should not be connected unless for structural reasons, e.g. 40 x 3 mm metal straps at 1.2 m centres, fixed at or just below ceiling level.
- Power point locations to be lined with 30 mm plasterboard and not positioned back to back (also consider switch positions).
- Any fire stops or cavity barriers should be flexible or only fixed to one frame.
- No restrictions for partition walls meeting separating wall.

Section 2: Separating floors for new building

Floors need to resist the passage of airborne sound or both airborne and impact sound. Three main options to achieve this are given (Figure 7.4). As with walls, example constructions, junction details and a checklist of items to consider are provided for guidance.

Floor type 1: Concrete base with soft covering

Constructions:

- Solid concrete in-situ slab, mass including any screed and/or ceiling finish, if used, 365 kg/m^2.

Concrete base with
soft covering

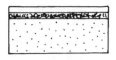

Concrete base with
floating layer

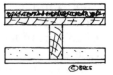

Timber base with
floating layer

Figure 7.4 Floor types.

- Solid concrete slab with permanent shuttering, mass including shuttering and screed and/or ceiling finish, if used, 365 kg/m^2.
- Concrete beams with infill blocks, mass of beams, blocks, screed and ceiling finish 365 kg/m^2.
- Concrete solid or hollow planks, mass of planks, screed and ceiling finish 365 kg/m^2.

With a soft covering to resist impact sound:

- Any resilient material, or material with a resilient base, minimum overall uncompressed thickness 4.5 mm.

Junctions:

- The mass of an external wall or cavity separating wall adjoining the floor should be at least 120 kg/m^2 (no requirement for an external wall with an area of openings exceeding 20%).
- The concrete floor base should pass through the wall leaf but not bridge any cavity.
- For an internal wall or solid separating wall with a mass less than 375 kg/m^2 the concrete floor base should pass through.

Consider:

- Fill all joints to floor construction to avoid air paths.
- Services (excluding gas pipes) penetrating a floor separating habitable rooms, above and below the floor, should be within an enclosure, mass 15 kg/m^2, incorporating 25 mm mineral wool.
- Reference Gas Safety (installation and use) Regulations 1998 for ventilation provisions to gas pipe ducts.

Floor type 2: Concrete base with floating layer

Constructions:

- Solid concrete in-situ slab, mass including any screed and/or ceiling finish, if used, 300 kg/m^2.

- Solid concrete slab with permanent shuttering, mass including shuttering and screed and/or ceiling finish, if used, 300 kg/m^2.
- Concrete beams with infill blocks, mass of beams, blocks, screed and ceiling finish 300 kg/m^2.
- Concrete solid or hollow planks, mass of planks, screed and ceiling finish 300 kg/m^2.

With a floating layer and resilient layer to resist impact sound:

- 18 mm tongued and grooved timber or wood-based boarding fixed to 45 x 45 mm battens on 25 mm mineral fibre, density 36 kg/m^3, resilient layer, *or*
- 65 mm reinforced cement/sand screed on resilient layer of 25 mm mineral fibre, density 36 kg/m^3 or 13 mm precompressed expanded polystyrene board (impact sound duty) or 5 mm extruded polyethylene foam (closed cell), density 30-45 kg/m^3.

Junctions:

- The mass of an external wall or cavity separating wall adjoining the floor should be at least 120 kg/m^2 (no requirement for an external wall with an area of openings exceeding 20%).
- The concrete floor base should pass through the wall leaf but not bridge any cavity.
- Turn resilient layer up at the edges and isolate skirtings from floating layer.
- For an internal wall or solid separating wall with a mass less than 375 kg/m^2 the concrete floor base should pass through.

Consider:

- For resistance to airborne sound *only* retain the full construction.
- Fill all joints to floor construction to avoid air paths.
- Services (excluding gas pipes) penetrating a floor separating habitable rooms, above and below the floor, should be within an enclosure, mass 15 kg/m^2, incorporating 25 mm mineral wool, isolated from floating layer.
- Reference Gas Safety (installation and use) Regulations 1998 for ventilation provisions to gas pipe ducts.

Floor type 3: Timber base with floating layer

Constructions:

- Floating layer of 18 mm tongued and grooved timber or wood-based boarding glued and spot-bonded to 19 mm plasterboard (or 24 mm cement-bonded particle board) on resilient layer of 25 mm mineral fibre, density 60–100 kg/m^3, on floor base of 12 mm timber or wood-based boarding nailed to structural timber joists. Ceiling of two layer, 30 mm thick plasterboard, joints staggered with absorbent material of 100 mm unfaced mineral fibre, density 10 kg/m^3.

- Floating layer of 18 mm tongued and grooved timber or wood-based boarding glued and spot-bonded to 19 mm plasterboard, nailed to 45 x 45 mm battens placed over joists (45 mm minimum width) with a resilient strip in between of 25 mm mineral fibre, density 80–140 kg/m^3. Ceiling of two-layer, 30 mm thick plasterboard, joints staggered with absorbent material of 100 mm unfaced mineral fibre, density 10 kg/m^3.
- Floating layer of 18 mm tongued and grooved timber or wood-based boarding glued, nailed or screwed to 45 x 45 mm battens placed over joists (45 mm minimum width) or between them with a resilient strip on the joists of 25 mm mineral fibre, density 80–140 kg/m^3. Ceiling of 6 mm plywood and two-layer, 25 mm thick plasterboard, joints staggered *or* 19 mm dense plaster on expanded metal lath, both to incorporate pugging (e.g. 60 mm limestone chips or 50 mm sand) of mass 80 kg/m^2 on a polyethylene liner.

Junctions:

- For a timber-frame wall, block air paths between the floor base and the wall and where joists are at right angles to wall; seal ceiling/wall-lining junction with tape or caulking.
- For a heavy masonry leaf, minimum mass 375 kg/m^2, joists may use any fixing method. Seal ceiling/wall-lining junction with tape or caulking.
- For a light masonry leaf, mass less than 375 kg/m^2, a free-standing isolation wall panel should be used. Joists may use any fixing method with ceiling taken through to masonry and sealed with tape or caulking.
- Turn resilient layer up at the edges and isolate skirtings from floating layer.

Consider:

- For resistance to airborne sound only the full construction should be retained.
- Control sound paths at floor perimeter and at service penetrations.
- Services (excluding gas pipes) penetrating a floor separating habitable rooms, above and below the floor, should be within an enclosure, mass 15 kg/m^2, incorporating 25 mm mineral wool and isolated from floating layer.
- Reference the Gas Safety (installation and use) Regulations 1998 for ventilation provisions to gas pipe ducts.

- Resilient layer to be of correct density and able to carry anticipated load.
- The floating layer and floor base *should not be bridged,* i.e. by fixings or services.

Section 3: Similar construction method for new building

A repetition of a previously tested and approved method of construction, for a wall, floor or stair, should be acceptable. The construction should be essentially similar with respect to such items as separating wall/floor constructions, size and shape of relevant rooms and arrangement of external wall openings. Minor differences are, however, allowed to:

- the construction of the inner or outer leaf of an external masonry cavity wall, subject to the inner leaf mass not being reduced;
- the specification of a floating layer on a concrete floor base; and
- the construction of a non-separating timber floor, e.g. in a maisonette.

The testing of existing construction, for the justification of its use for new build, should be in accordance with BS 2750: Part 4: 1980 (where the tests determine the standardized level differences for airborne sound insulation) and Part 7: 1980 (where the tests determine the standardized impact sound pressure levels). The test report obtained for a particular construction should contain the following information:

- organization conducting the test, including UKAS accreditation number if applicable;
- name of person in charge of test;
- date of test;
- address of building tested;
- brief details of test, including equipment and procedures;
- description of building, including room dimensions and mass of elements;
- test results, in tabular and graphical form.

For full test methods, result assessments and sound insulation values reference should be made to the Approved Document, paragraphs 3.5–3.8.

Section 4: Test chamber evaluation for new construction

An approved type of test chamber can be used to assess a proposed separating/flanking wall construction where the sound insulation between at least two pairs of rooms is not less than a prescribed value. The test report can be regarded as evidence of compliance for a specific type of construction, where only the following may be changed:

- separating wall and flank wall dimensions;
- flank wall door and/or window positions; and
- attachment of partition walls to the separating wall.

The test report obtained for a particular construction should contain the following information:

- organization(s) operating the test chamber and conducting the acoustic measurements, including UKAS accreditation number(s) if applicable;
- date of test;
- description of test chamber;
- brief details of test, including equipment and procedures;
- full details of materials and test construction, including room dimensions and mass of elements;
- test results, where the modified weighted standardized level difference value obtained from each measurement should not be less than 55 dB.

For test procedure and sound insulation value requirements, reference should be made to the Approved Document, paragraphs 4.3–4.5.

Section 5: Remedial work in conversions

The enclosing construction of an existing building converted into separate dwellings may inherently offer an acceptable level of sound insulation. This would be the case if the mass of the construction is within 15% of constructions specified in Sections 1 and 2, or compliance could be shown with the test requirements of Section 6. Where this cannot be achieved the level of sound insulation should be improved by the provision of the following wall treatment, one of the floor treatments and the stair treatment (Figure 7.5). In the upgrading of existing construction consideration may need to be given to the suitability of load-bearing elements, e.g. floor joists supporting heavy pugging or plasterboard ceilings.

Wall treatment 1: Independent lining and absorbent cavity material

Construction:

- **Independent** lining of 25 mm plasterboard, two layers with joints staggered (minimum 25 mm from existing wall), on framing (minimum 13 mm from existing wall) or 30 mm plasterboard if no frame

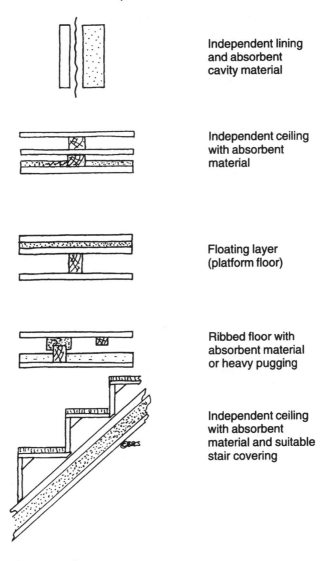

Independent lining
and absorbent
cavity material

Independent ceiling
with absorbent
material

Floating layer
(platform floor)

Ribbed floor with
absorbent material
or heavy pugging

Independent ceiling
with absorbent
material and suitable
stair covering

Figure 7.5 Treatment for conversions.

is used, with an absorbent material of 25 mm thick mineral wool, density 10 kg/m^3. Seal perimeter of lining with tape or mastic.

Consider:

- Provide independent lining to *both* faces of existing wall unless it is at least 100 mm thick masonry, plastered on both sides.
- The absorbent material *only* may bridge the cavity but it should not be tightly compressed in place.

- Ceiling/lining junction to be sealed with tape or caulking.
- Isolate skirtings from floor and floating layer.

Floor treatment 1: Independent ceiling with absorbent material

Construction:

- Seal existing floor boarding with caulking or overlay with hard-board.
- Replace or upgrade existing ceiling to 30 mm of plasterboard in two or three layers, joints staggered, unless existing ceiling is lath and plaster.
- New independent ceiling of 30 mm plasterboard, two layers with joints staggered (minimum 100 mm from existing ceiling), fixed to new independent ceiling joists (minimum 25 mm from existing ceiling) with an absorbent material in between of 100 mm thick mineral wool, density 10 kg/m^3. Seal ceiling perimeter with tape or mastic.

Consider:

- For a high window location the independent ceiling depth may be reduced to allow for a pelmet recess.
- Penetrations by piped services, including gas pipes, follow the guidance as described for new build.

Floor treatment 2: Floating layer (platform floor)

Construction:

- Replace or upgrade existing ceiling to 30 mm of plasterboard in two or three layers, joints staggered, unless existing ceiling is lath and plaster.
- Provide 100 mm mineral wool absorbent layer if possible.
- Floating layer of 18 mm tongued and grooved timber or wood-based boarding glued and spot-bonded to 19 mm plasterboard or one- or two-layer material with a mass of 25 kg/m^2, all joints glued. Both on resilient layer of 25 mm mineral fibre, density 60–100 kg/m^3.

Consider:

- Seal any new ceiling perimeter with tape or caulking.
- Resilient layer to be of correct density and able to carry anticipated load.
- Isolate skirtings from floor and floating layer.

- Penetrations by piped services, including gas pipes, follow the guidance as described for new build.

Floor treatment 3: Ribbed floor with absorbent material or heavy pugging

Construction:

- Replace or upgrade existing ceiling to 30 mm of plasterboard in two or three layers, joints staggered, unless existing ceiling is lath and plaster. Provide 100 mm mineral wool absorbent layer if possible.
- Floating layer of 18 mm tongued and grooved timber or wood-based boarding glued and spot-bonded to 19 mm plasterboard or one- or two-layer material with a mass of 25 kg/m^2, all joints glued, nailed or screwed to 45 x 45 mm battens placed over joists (45 mm) or between them with a resilient strip on the joists of 25 mm mineral fibre, density 80–140 kg/m^3. With 100 mm mineral fibre absorbent layer between joists, density 10 kg/m^3.
- Pugging (e.g. 60 mm limestone chips or 50 mm sand) of mass 80 kg/m^2, on a polyethylene liner, may be used as an alternative.

Consider:

- Seal any new ceiling perimeter with tape or caulking.
- Resilient layer to be of correct density and able to carry anticipated load.
- Isolate skirtings from floor and floating layer.
- Owing to the removal of the existing floorboarding, additional strutting may be necessary for floor stability.
- Penetrations by piped services, including gas pipes, follow the guidance as described for new build.
- Where the utilization of **floor treatments 1, 2 or 3** is **not practical** two alternative solutions may be adopted.

Floor treatment 4: Alternative independent ceiling with absorbent material

This is a less onerous version of treatment 1, where the joists of the independent ceiling can be located between the existing joists, thus allowing less reduction in ceiling height (although sound insulation value is lower).

Floor treatment 5: Alternative floating layer (platform floor)

As a less onerous version of treatment 2 the resilient layer specification is reduced, resulting in a lower floor level (although sound insulation value is lower).

Stair treatment 1: Independent ceiling with absorbent material and suitable stair covering

Construction:

- Lay a soft covering minimum 6 mm thick, e.g. carpet, over stair treads.
- Construct an independent ceiling below the stair as described for **floor treatment 1**, or
- Where a cupboard is located below the stairs adopt an enclosure utilizing two layers of 12.5 mm plasterboard (or equivalent mass), use a small, heavy, well-fitting access door and line the underside of the stairs within the cupboard with 12.5 mm plasterboard, incorporating 25 mm mineral wool.

Section 6: Field and laboratory tests for conversions

An existing construction method may be repeated where this method has been previously built and tested. Field or laboratory tests may be used that use a **base** wall or floor construction as a starting point. Specific reference is made in the Approved Document to the various parts of BS 2750: *Measurement of sound insulation in buildings and of building elements*, and additional rules, which should be followed, are stated in paragraphs 6.4–6.12 (Figure 7.6).

For tests of remedial treatments applied to separating walls, floors and stairs and for laboratory tests of remedial treatments the test report obtained for a particular construction should contain the following information:

- organization conducting the test, including UKAS accreditation number if applicable;
- name of person in charge of test;
- date of test;
- brief details of test, including equipment and procedures;
- description of treatment tested or construction tested;
- test results, in tabular and graphical form.

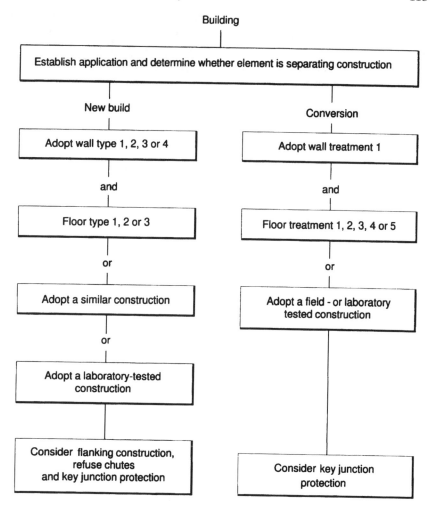

Figure 7.6 Summary of sound resistance requirements.

Appendix A: Method for calculating mass

The Approved Document concludes with a calculation method, which can be used as an option to the direct use of manufacturers' figures, those stated in a current National or European Certificate (European Technical Approval). For a given coordinating height of masonry course the following formulae are stated:

$M = T(0.79D + 380)$ for 75 mm masonry course
$M = T(0.86D + 255)$ for 100 mm
$M = T(0.92D + 145)$ for 150 mm
$M = T(0.93D + 125)$ for 200 mm

where M = mass of 1 m^2 of leaf in kg/m^2; T = thickness of masonry in metres; D = density of masonry units in kg/m^2 (at 3% moisture content); N = number of finished faces; P = mass of 1 m2 of wall finish in kg/m^2, e.g. cement render 29, gypsum plaster 17, lightweight plaster and plasterboard 10.

CHAPTER 8

Approved Document F: Ventilation

Requirements F1 and F2, and the guidance contained in the Approved Document, seek to control air quality in buildings and the limitation of condensation in roof voids. With a background of workplace ventilation requirements and cases of legionnaires' disease, ventilation provisions are now applied to non-domestic buildings above and beyond those previously covering just sanitary accommodation. The need for greater control of condensation has also arisen out of the need to provide ever-increasing levels of thermal insulation.

REQUIREMENT F1: MEANS OF VENTILATION

There shall be adequate means of ventilation provided for people in the building.
Requirement F1 does not apply to a building or space within a building:

(a) into which people do not normally go; or
(b) which is used solely for storage; or
(c) which is a garage used solely in connection with a single dwelling.

Without satisfactory ventilation a health hazard may arise from the accumulation of moisture, leading to mould growth, or from pollutants emanating from within the building. In an effort to avoid these situations the following ventilation principles need to be applied.

- Extract water vapour at source where it is produced in significant quantities: e.g. kitchens, utility rooms and bathrooms.
- Extract pollutants at source where they are produced in significant quantities: e.g. contaminants originating from harmful processes or smoking rooms.
- Rapidly dilute pollutants and water vapour originating within

habitable rooms, occupiable rooms and sanitary accommodation where necessary.

- Disperse residual water vapour in the long term by a minimum background supply of fresh air, which should not affect comfort, security or allow rain penetration.
- Non-domestic mechanical ventilation or air-conditioning systems should provide the ventilation provisions listed above and be designed, installed and commissioned so as not to be detrimental to health, and allow access for maintenance.

A series of definitions in relation to Requirement F1 is listed:

Ventilation opening – includes any permanent or closable means of ventilation opening directly to the external air: e.g. openable parts of a window, a louvre, airbrick, progressively openable ventilator, trickle ventilator or an external door.

Habitable room – used for dwelling purposes but not solely a kitchen.

Bathroom – including a shower room, may include sanitary accommodation.

Occupiable room – in a non-domestic building includes an office, workroom, classroom, hotel bedroom, etc., but not spaces or rooms principally used for circulation, plant or storage areas, bathrooms or utility rooms.

Domestic buildings – includes houses, flats, student accommodation and residential homes.

Non-domestic buildings – means all other purpose groups including hotels.

Passive stack ventilation (PSV) – utilizes the natural stack effect of different internal and external air temperatures and wind pressures over the roof to ventilate rooms via ceiling grilles, ducts and roof terminals. Reference BRE Information Paper 13/94.

For **rapid ventilation** utilize an opening window, 1.75 m above floor level.

For **background ventilation** utilize trickle ventilators, air bricks (with hit-and-miss grilles), or suitably designed opening windows, 1.75 m above floor level, as illustrated in Diagram 1 and 4 of the Approved Document.

For **extract ventilation** utilize a PSV (for domestic facilities) or mechanical system, both operated manually and/or automatically by sensor or controller or an appropriate open-flued appliance (for domestic buildings only).

Section 1: Domestic buildings

To show compliance with the established ventilation principles reference should be made to Figure 8.1, which outlines the provisions to be applied for specific rooms within domestic buildings.

To supplement the guidance contained in Figure 8.1 consideration should be given to the following additional matters.

- Where kitchens, utility rooms, bathrooms and sanitary accommodation contain no openable windows then provide: mechanical extract rated as indicated in Figure 8.1 with a 15 min overrun (e.g. control via the light switch); or PSV; or an appropriate open-flued appliance (*not in bathrooms*); all with an air inlet, e.g. a 10 mm gap under the door.
- As an **alternative approach** to the background ventilation illustrated in Figure 8.1 adopt an average of 6000 mm^2 per specified room, with a minimum 4000 mm^2 to each room.
- For combined-use rooms, e.g. a kitchen/diner, adopt the most stringent provision only.

The installation of open-flued appliances together with mechanical extract ventilation can cause a serious threat to life with respect to the spillage of flue gases. These gases can be drawn back into the room even if the fan is not located in the same room. The appliance therefore needs to operate safely whether the fan is running or not, and guidance should be taken from the following.

- **Gas appliances** located in a kitchen would require the fan to operate at a maximum extract rate of 20 l/s. A spillage test should be carried out, in accordance with BS 5440: Part 1, Clause 4.3.2.3,

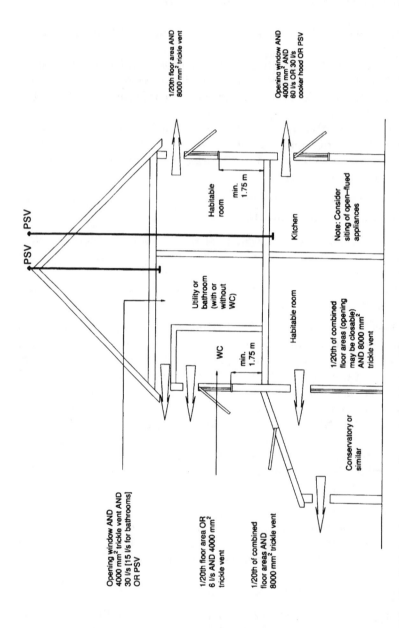

Figure 8.1 Ventilation of rooms in domestic buildings. To treat two rooms as a single room for ventilation purposes a minimum permanent opening of 1/20th of the combined floor area should be provided between the rooms.

whether the fan is in the same room or not. For further reference, see BRE Information Paper 21/92.

- **Oil-fired appliances** should be installed in accordance with Technical Information Note T1/112, obtained from the Oil Firing Technical Association for the Petroleum Industry (OFTEC).
- **Solid fuel appliances** should not have mechanical extract ventilation provided in the same room; contact Heating Equipment Testing and Approval Scheme (HETAS).

As an **alternative approach** to the provisions illustrated in Figure 8.1 reference can be made to one of the following:

- BS 5925: 1991 Code of practice for ventilation principles and designing for natural ventilation;
- BS 5720: 1979 Code of practice for mechanical ventilation and air-conditioning in buildings;
- BS 5250: 1989 Code of practice for the control of condensation in buildings;
- BRE Digest 398 Continuous mechanical ventilation in dwellings: design, installation and operation.
- Reference can also be made to Approved Document – Basements for dwellings.

Section 2: Non-domestic buildings

To show compliance with the established ventilation principles for non- domestic buildings reference should first be made to Figure 8.2 and then to the provisions outlined for specialist activities, car parks and mechanical ventilation/air-conditioning plant.

To supplement the guidance contained in Figure 8.2 consideration should be given to the following additional matters.

- Where kitchens, bathrooms and sanitary accommodation contain no openable windows then provide mechanical extract rated as indicated in Figure 8.2 with a 15 min overrun (e.g. control via the light switch or occupant detecting sensor) with an air inlet, e.g. a 10 mm gap under the door.
- Where an occupiable, **non-smoking**, room utilizes mechanical ventilation it should be rated at a minimum of 8 l/s of fresh air per occupant. This should be increased to 16 l/s for rooms used for **light smoking**.
- For **common spaces** (not just used for circulation) where large numbers of people may gather, e.g. a shopping mall or foyer, provide: openings for natural ventilation, minimum 1/50th of space floor area; or a mechanical ventilation system designed to supply fresh air at a minimum 1 l/s per m^2 of space floor area.

As an **alternative approach** to the provisions illustrated in Figure 8.2 reference can be made to one of the following:

- BS 5925: 1991 Code of practice for ventilation principles and designing for natural ventilation;
- The Chartered Institution of Building Services Engineers (CIBSE) Guide A: *Design data* and CIBSE Guide B: *Installation and equipment data*.

Specialist activities may warrant the application of specific design guidance. The Approved Document highlights the following.

- **School and other educational establishments**. The provisions illustrated in Figure 8.2 can be used (subject to six air changes per hour for sanitary accommodation) or reference can be made to the Education (School Premises) Regulations including DFEE Building Bulletin 87, Guidelines for environmental design in Schools and Design Note 29 if fume cupboards are deemed necessary.
- **Workplaces**, including work processes, should follow the Health and Safety Executive (HSE) Guidance Note EH22 *Ventilation of the workplace*.
- **Hospitals**. Ventilation requirements for the large range of room uses found in hospital premises are given in the *Department of Health Activity Data Base*; guidance and provisions are then given in Department of Health Building Notes, e.g. HBN 4 *Adult Acute Wards*, HBN 21 *Maternity Departments* and HBN 46 *General medical practice premises*.
- **Building services plant rooms**. Emergency ventilation provisions are necessary to control contaminating gas releases, e.g. refrigerant leak: reference HSE Guidance Note EH22 *Ventilation of the workplace* and BS 4434: 1989 *Specification for safety aspects in the design, construction and installation of refrigeration appliances and systems*.
- **Restrooms where smoking is permitted**. Under the Workplace Regulations non-smokers in these occupiable rooms should be protected from the discomfort caused by tobacco smoke by natural ventilation in accordance with Figure 8.2 and mechanical ventilation, minimum 16 l/s per occupant.
- **Commercial kitchens** should follow the guidance given in CIBSE Guide B, Tables B2.3 and B2.11.

The ventilation of **car parks** relates to the guidance contained in Section 11 of Approved Document B, which should also be referred to. **Naturally ventilated car parks** are deemed satisfactory where they can be regarded as **open sided** with permanent openings a minimum of 5% of the floor area at each level (at least 2.5% of which should be in opposing walls). For **mechanically ventilated car parks** adopt a minimum three air changes per hour *and* permanent openings (2.5% of floor

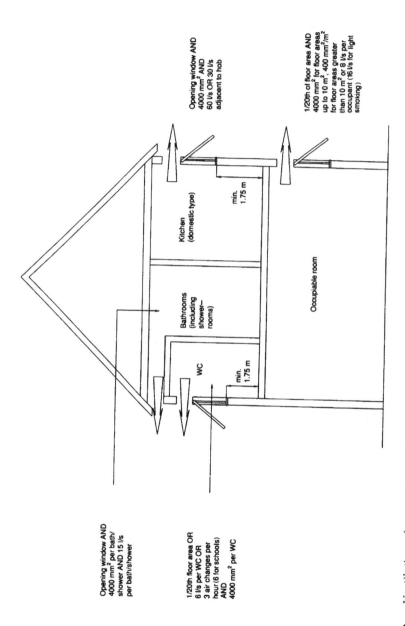

Opening window AND
4000 mm² per bath/
shower AND 15 l/s
per bath/shower

1/20th floor area OR
6 l/s per WC OR
3 air changes per
hour (6 for schools)
AND
4000 mm² per WC

Opening window AND
4000 mm² AND
60 l/s OR 30 l/s
adjacent to hob

1/20th of floor area AND
4000 mm² for floor areas
up to 10 m², 400 mm²/m²
for floor areas greater
than 10 m² or 8 l/s per
occupant (16 l/s for light
smoking)

Kitchen
(domestic type)

min.
1.75 m

Bathrooms
(including
shower—
rooms)

WC

min.
1.75 m

Occupiable room

Figure 8.2. Ventilation of rooms in non-domestic buildings. As an alternative to a mechanical extract fan a PSV can be used for domestic-type facilities.

area); or six air changes per hour for basement car parks and a local rate of ten air changes per hour where cars queue inside the building at exit and ramp locations.

An **alternative approach** for car parks is to limit concentrations of carbon monoxide to 50 parts per million (averaged over an 8 hour period) and 100 parts per million for ramp and exit type locations (15 min period). For further guidance refer to the Association for Petroleum and Explosives Administration *Code of practice for ground floor, multi-storey and underground car parks* and the CIBSE Guide B, Section B2-6 and Table B2-7.

In the **design of mechanical ventilation/air-conditioning plant**, fresh air supply inlets should be located to avoid drawing in excessively contaminated air: examples include the close location of a flue, evaporative cooling tower, vehicle turning area, etc. With respect to legionella contamination reference should be made to HSE *The control of legionellosis including legionnaires' disease*, and for guidance on recirculated air refer to HSE Workplace (Health, Safety and Welfare) Regulations 1992 *Approved Code of Practice and guidance* L24. As an **alternative approach** follow:

- BS 5720: 1979 Code of practice for mechanical ventilation and air-conditioning in buildings;
- CIBSE Guide B: Installation and equipment data.

Two important concluding matters relating to **mechanical ventilation/ air-conditioning plant** are access for maintenance and commissioning of the system.

- Provide access to replace filters and to clean ductwork.
- In a central plant room allow 600 mm width for access to plant (2000 mm headroom) and 1100 mm where maintenance operations are carried out. Also refer to Building Services Research and Information Association (BSRIA) Technical Note 10/92: *Space allowances for building services distribution systems.*
- For systems serving floor areas in excess of 200 m^2 and undertaken in accordance with the guidance outlined above, test reports and commissioning certificates should be provided to the building control authority, e.g. CIBSE commissioning codes confirming system performance accords with approved specification.

REQUIREMENT F2: CONDENSATION IN ROOFS

Adequate provision shall be made to prevent excessive condensation:

(a) in a roof; or

(b) in a roof void above an insulated ceiling.

The reasons for controlling excessive condensation in roof voids are that it can substantially and permanently reduce the performance of the roof thermal insulation material and the roof structure. Before applying the approved document guidance consideration should first be given to the following.

The reasons for controlling excessive condensation in roof voids are that it can substantially and permanently reduce the performance of the roof thermal insulation material and the roof structure. Before applying the approved document guidance consideration should first be given to the following:

- The ventilation provisions outlined in the approved document only relate to **cold roof** constructions, where moisture can permeate the insulation, and not **warm roof** decks, since these should not allow permeation of moisture from within the building.
- Small roofs over porches and bay windows may not require ventilation, owing to their limited risk to health and safety.
- The provisions apply to a roof of any pitch, including a wall at 70° or more.
- Ventilation openings may be in a continuous or intermittent form with a screen, baffle, etc. to prevent the entry of insects, etc.
- Purpose-made components can be used to stop insulation materials restricting air flow paths at eaves and other locations.
- The 1994 edition of BRE Report 262 *Thermal insulation: avoiding risks* contains a wealth of supporting **detailed guidance**, and should be referred to where necessary.

Section 1: Roofs with a pitch of 15° or more (pitched roofs)

The fundamental principle is to provide adequate cross-ventilation of the roof void, above the insulated ceiling, to remove moisture-laden air. This can be achieved as follows.

- **Pitched roofs** should have eaves-level vents at opposite sides of the roof so as to provide the equivalent of a continuous 10 mm wide opening.

- **Lean-to** (or **mono-pitch**) **roofs** should have the same eaves-level vent and the equivalent of a continuous 5 mm wide opening at high level, e.g. venting tiles or abutment ventilation strip.
- The **alternative approach** is to follow BS 5250: 1989 Code of practice for the control of condensation in buildings.

Section 2: Roofs with a pitch of less than 15° and those where the ceiling follows the pitch of the roof

As with Section 1, cross-ventilation arrangements are needed, which must take into account the reduced roof void area and restriction of air flow paths to sloping ceilings following the pitch of the roof, e.g. a loft room. This can be achieved as follows.

- **Flat roofs** should have eaves-level vents at opposite sides of the roof so as to provide the equivalent of a continuous 25 mm wide opening. A minimum 50 mm clear air space should be retained between the underside of the roof deck and the insulation, utilizing counter-battens where necessary. If the roof plan is not a simple rectangle or the span exceeds 10 m, adopt aggregate ventilation openings of 0.6% of roof plan area.
- **Roofs with ceilings following pitch** should have the same eaves-level vent and a continuous 5 mm wide opening at ridge level, e.g. ridge venting tiles. A minimum 50 mm clear air space should be retained between the underside of the roofing felt and the insulation.
- Where one side of the roof **abuts** an external wall, an abutment ventilation strip, vent cowl(s) or venting tiles may be necessary to preserve cross-ventilation where possible.
- Note that a **vapour check** (or vapour control layer), e.g. foil-backed plasterboard, may help to reduce moisture levels into the roof void, but it should not be regarded as a vapour barrier, and the cross-ventilation provisions should therefore be retained.
- The **alternative approach** is to follow BS 5250: 1989 Code of practice for the control of condensation in buildings.

CHAPTER 9

Approved Document G: Hygiene

Under the heading of hygiene, this Approved Document brings together guidance to show compliance with Requirements G1, G2 and G3. These relate to sanitary accommodation, washing facilities and the installation of unvented hot water storage systems where the submission of a building notice or full plans is necessary to accord with Regulations 11(1), 12(4) and 13(3).

REQUIREMENT G1: SANITARY CONVENIENCES AND WASHING FACILITIES

1. Adequate sanitary conveniences shall be provided in rooms provided for that purpose, or in bathrooms. Any such room or bathroom shall be separated from places where food is prepared.
2. Adequate washbasins shall be provided in:
   (a) rooms containing water closets; or
   (b) rooms or spaces adjacent to rooms containing water closets. Any such room or space shall be separated from places where food is prepared.
3. There shall be a suitable installation for the provision of hot and cold water to washbasins provided in accordance with paragraph (2).
4. Sanitary conveniences and washbasins to which this paragraph applies shall be designed and installed so as to allow effective cleaning.

REQUIREMENT G2: BATHROOMS

A bathroom shall be provided containing either a fixed bath or shower bath, and there shall be a suitable installation for the provision of hot and cold water to the bath or shower bath.
Requirement G2 applies only to dwellings.

Section 1: Sanitary conveniences and washing facilities

The wording of Requirement G1 is self-explanatory, seeking to require satisfactory and sufficient toilet accommodation with associated washing provisions to *all* building purpose groups. A summary of the Approved Document guidance is as follows.

- A dwelling, including a house, flat, maisonette or house in multi-occupation, should have a minimum of one WC and one washbasin.
- Only a door, not a lobby, is needed to separate a WC or urinal from a food preparation area, including a kitchen or washing up area.
- The washbasin should be located in the same room as the WC or in an adjoining room or area, subject to this not being used for food preparation. The washbasin should have a hot water supply (either direct from a central source or an instantaneous water heater) and discharge via a trap to the foul water drainage system.
- A flushing WC or urinal should cleanse the bowl effectively and discharge via a trap to the foul water drainage system.
- A WC, urinal or washbasin should have a smooth, non-absorbent and easy-clean internal surface.
- A WC utilizing a macerator, pump and small-bore pipe can discharge to the foul water drainage subject to the system holding a current Agrément Certificate (or equivalent European Technical Approval) *and* the availability of a traditional WC connected by gravity to the foul water drainage system.
- Chemical WCs and urinals may *only* be used where a suitable water supply and the means to dispose of foul water is not available.

For the number, type and siting of appliances, in buildings other than dwellings and in addition to the above, reference should also be made to:

- Offices, Shops and Railway Premises Act 1963;
- Factories Act 1961;
- Food Hygiene (General) Regulations 1970;
- Approved Document M – Access and facilities for disabled people.

As an **alternative approach** to the above guidance, reference can also be made to BS 6465: Part 1: 1984 *Code of Practice for scale of provision, selection and installation of sanitary appliances.* This British Standard incorporates minimum requirement tables for a range of building purpose groups, including dwellings, residential homes, offices and shops, factories, schools, assembly buildings and hotels.

Section 2: Bathrooms

The Approved Document guidance is straightforward, reflecting the self-explanatory wording of the requirement. It is important to note that bathroom provisions apply *only* to a **dwelling**, and reference will therefore need to be made to the legislation listed under Requirement G1 concerning bath and shower facilities in offices and shops, factories, hotels, etc.

- A dwelling, including a house, flat, maisonette or house in multi-occupation should have a minimum of one bathroom with a fixed bath or shower.

- The bath or shower should have a hot water supply (either direct from a central source or an instantaneous water heater) and discharge via a trap to the foul water drainage system.
- A bath or shower utilizing a macerator, pump and small-bore pipe can discharge to the foul water drainage subject to the system holding a current Agrément Certificate (or equivalent European Technical Approval).

REQUIREMENT G3: HOT WATER STORAGE

A hot water storage system that has a hot water storage vessel which does not incorporate a vent pipe to the atmosphere shall be installed by a person competent to do so, and there shall be precautions:

(a) to prevent the temperature of stored water at any time exceeding 100 °C; and
(b) to ensure that the hot water discharged from safety devices is safely conveyed to where it is visible but will not cause danger to persons in or about the building.

Requirement G3 does not apply to:

(a) a hot water storage system that has a storage vessel with a capacity of 15 litres or less;
(b) a system providing space heating only;
(c) a system which heats or stores water for the purposes only of an industrial process.

The utilization of sealed hot water storage vessels within heating systems gives rise to potential health risks to persons in or about buildings. The vessel itself is subject to high internal pressures, and water discharged via safety devices will be very hot. The guidance is split into two sections dependent on whether the vessel is up to or over 500 litres capacity. The Water Supply (Water Fittings) Regulations 1999 should also be considered.

Section 3: Systems up to 500 litres and 45 kW

An unvented hot water storage system should be installed by a **competent person**, i.e. a Registered Operative holding a current identity card issued by one of the following:

- Construction Industry Training Board;
- Institute of Plumbing;
- Association of Installers of Unvented Hot Water Systems (Scotland and Northern Ireland);
- BBA Approved Installer;
- an equivalent body.

The design of the system itself should take the form of a **proprietary unit or package**, should incorporate at least two factory-fitted temperature- activated safety devices operating in sequence, the specification of which is dependent on whether the heating system is direct or indirect, and should be approved by:

- a member body of the European Organization for Technical Approvals (EOTA); or
- a certification body having National Accreditation Council for Certification Bodies' (NACCB) accreditation, e.g. to BS 7206; or
- an equivalent independent assessment.

For installations with an EOTA or NACCB approval, inspection by the controlling authority is unlikely to be necessary. The typical discharge pipe arrangements from a safety device should include a tundish, for visual warning, within 500 mm of the device and its termination to a safe location, e.g. a trapped gulley, avoiding asphalt or felt roofs and non-metallic rainwater goods. For specific design guidance reference should be made to paragraph 3.9, Diagram 1 and Table 1 of the Approved Document. The **alternative approach** is to refer to BS 6700: 1987.

Section 4: Systems over 500 litres or over 45 kW

These systems will generally be designed on an individual basis where EOTA or NACCB approvals would not be appropriate. The system should therefore be designed by a suitably qualified engineer with a **competent person** used to install the system. The provision of safety devices depends on whether the system has a power input of not more than 45 kW, where reference should be made to the relevant recommendations of BS 6700: 1987 and BS 6283: Parts 2 and 3: 1991. Discharge pipes should accord with Section 3.

CHAPTER 10

Approved Document H: Drainage and waste disposal

The satisfactory drainage of a building and the efficient disposal of refuse are both matters that could have a dramatic influence on public health and safety. A blocked foul or surface water drain, for example, could cause effluent leakage or flooding; a badly designed septic tank could allow leakage into the subsoil or watercourse; and insufficient dustbin provision could give rise to rodent infestation. It is therefore important that compliance with the functional requirements, H1–H4, is shown by referring to the guidance contained in Approved Document H.

REQUIREMENT H1: FOUL WATER DRAINAGE

1. Any system which carries foul water from appliances within the building to a sewer, a cesspool or a septic or settlement tank shall be adequate.
2. 'Foul water' in sub-paragraph (1) means waste water which comprises or includes:
   (a) waste from a sanitary convenience or other soil appliance,
   (b) water which has been used for cooking or washing.

To comply with the requirement the foul water drainage system should:

- convey the foul water flow to a suitable outfall;
- minimize the risk of leakage and blockages;
- prevent foul air from entering the building;
- be ventilated and made accessible for blockage clearance.

An important point to note is that the requirement seeks to control the foul water drainage system *only* and *not* the suitability of the outfall itself, where guidance from the local water authority should be sought.

Table 10.1 Flow rates

Number of dwellings (1 WC, 1 bath, 1 sink and 1 or 2 washbasins)	Flow rate (l/s)	Individual appliances (l/s)	Flow rate (l/s)
1	2.5	Washdown WC	2.3
5	3.5	Urinal (per person unit)	0.15
10	4.1	Washbasin (32 mm branch)	0.6
15	4.6	Sink (40 mm branch)	0.9
20	5.1	Bath (40 mm branch)	1.1
25	5.4	Auto washing machine	0.7
30	5.8	Shower	0.1
		Spray tap basin	0.06

Before outlining the guidance contained in Sections 1 and 2 of the Approved Document flow rate data may be necessary to verify that the capacity of the system will be sufficient for the expected flow at any point in the system. Reference can therefore be made to Table 10.1.

Section 1: Sanitary pipework

Figure 10.1 illustrates the principle provisions for sanitary pipework. The specific provisions are then discussed under each element of the above ground drainage system starting with the need to provide a water seal (or trap) at all points of discharge into the system.

Traps

- Should prevent foul air in the system entering the building by retaining, under working/test conditions, a minimum seal of 25 mm.
- Should be removable, have a cleaning eye or come away with the appliance.

Branch discharge pipes

- Should discharge to another branch discharge pipe or a discharge stack, unless the appliances are on the ground floor, where they may discharge to a stub stack, gulley (waste water only) or directly to the drain.
- Should not discharge into a stack lower than 450 mm above the drain invert level, for single dwellings up to three storeys.
- Should not discharge into a stack lower than 750 mm above the drain invert level, for buildings up to five storeys. For buildings

over five storeys [and those over 20 storeys] connect ground [ground and first] floor appliances to a separate stack (gulley or drain if applicable).

- Ground-floor WCs may discharge directly to the drain where the maximum dimension above drain invert level is 1.5 m.
- Pipe entry into a gulley should be above the water seal but below the grating or access plate.
- Avoid bends or utilize the largest radius possible.
- If the length and slope exceeds the figures given in Figure 10.1 then to avoid loss of trap seals due to system pressures the branch pipe should be ventilated to the external air by a branch ventilation pipe,

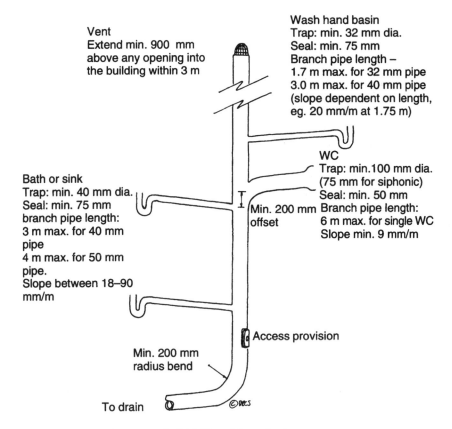

Figure 10.1 Discharge stack (S&VP) and branch pipes.

Note: For unvented (common) branch discharge pipes serving more than one appliance apply the following:

WCs – up to eight in number or 15 m branch length, 100 mm pipe size
Bowl urinals – up to five (branch as short as possible), 50 mm pipe size
Stall urinals – up to seven (branch as short as possible), 65 mm pipe/trap size
Washbasins – up to four or 4 m branch length (no bends), 50 mm pipe size.

or via the discharge stack (modified single stack system) or via a ventilating stack (ventilated system).

- Branch ventilating pipes should: connect within 300 mm of the trap and above spillover level (of the highest appliance) at stack; terminate as discharge stack (Figure 10.1) and be a minimum 25 mm diameter (35 mm diameter if longer than 15 m or pipe has more than five bends).
- Unless removable traps (or appliances) can be used then lengths of pipe should have rodding access.

Discharge stacks

- Minimum diameters: 50 mm (no WCs) up to 1.2 l/s 65 mm (no WCs) up to 2.1 l/s 75 mm (one siphonic WC) up to 3.4 l/s 90 mm up to 5.3 l/s 100 mm up to 7.2 l/s.
- Offsets in the 'wet' portion should be avoided, but they may be used in buildings up to three storeys if no branch connection is made within 750 mm. In higher buildings a separate ventilating stack may be needed with connections above and below offset.
- Should be inside the building if more than three storeys high.
- To avoid loss of trap seals due to system pressures the stack should be ventilated to the external air, terminated as in Figure 10.1. This may be reduced to 75 mm diameter for houses up to two storeys, minimum 50 mm in other cases if the drain is liable to surcharging. Consideration could also be given to the use of an **air admittance valve** located within the building and subject to a current Agrément Certificate.
- Allow reasonable access provisions for rodding purposes and repair of system pipework.

Stub stacks

- An unventilated stub stack may connect above ground to a ventilated discharge stack or to a drain not subject to surcharging.
- No branch should connect more than 2 m above the invert level of the drain (or discharge stack connection), 1.5 m maximum for a WC.
- Length of branch drain serving a stub stack should be a maximum of 6 m for a single appliance and 12 m for a group of appliances (unless ventilated).

A range of materials may be used for sanitary pipework, including cast iron, copper, galvanized steel, uPVC, polypropylene and plastics. Positive fixing methods should be adopted that allow for thermal movement of the pipes. Consideration may also need to be given to the separation of different metals to avoid electrolytic corrosion. Once installed, the pipes, fittings and joints should withstand an air test (or smoke test, but not suitable for uPVC pipes) of positive pressure, minimum 38 mm water gauge, for at least 3 min; traps to retain a minimum 25 mm seal.

An **alternative approach** to the guidance above is to refer to BS 5572: 1978 *Code of practice for sanitary pipework*. BRE Digests 248 *Sanitary Pipework: Part 1 Design basis* and 249 *Sanitary Pipework: Part 2 Design of pipework*, are also a useful reference source. This guidance is also more appropriate for large buildings.

Section 2: Foul drainage

The below-ground foul water drainage system comprises the necessary pipes and fittings to connect sanitary pipework (i.e. discharge stacks, stub stacks and gullies) to the outfall (i.e. a public or private sewer, cesspool, septic tank or settlement tank). Foul water drains may discharge to a combined public sewer, which carries both foul and surface water, although the pipe sizes may need to be increased to take account of both flow rates. The system layout should be kept as simple as possible, should minimize changes of direction and gradient, and should provide for access points where blockages could not be cleared without them. The specific provisions are discussed under headings reflecting the components of the system.

Pipes

- The drainage pipes should have sufficient capacity to carry the anticipated flow and be laid to falls. Table 10.2 gives specimen values.
- Any change of gradient should incorporate an access point.
- Pipes should be laid in straight lines, or slight curves if blockages can still be cleared. Bends used should be close to an inspection chamber or manhole and to the foot of discharge or ventilating stacks.
- The system should be ventilated at or near the head of each main drain, any branch drain longer than 6 m serving a single appliance or 12 m serving a group of appliances, and to a drain fitted with an intercepting trap. A ventilating discharge stack would achieve this.
- Where pipes run under or near a building, on piles/ground beams, in common trenches or in unstable ground, precautions may be necessary to avoid settlement of drain runs.
- For drain runs under large buildings adopt a minimum 100 mm granular or other flexible filling around the pipe. Where excessive

Table 10.2 Minimum gradients and maximum capacities for foul water drains

Pipe diameter (mm)	Gradient	Flow rate (l/s)
100	1:20	13.0*
100	1:40	9.2 (at peak flow rate < 1 l/s)
100	1:60	7.4*
100	1:80 (min. 1 WC)	6.3 (at peak flow rate > 1 l/s)
150	1:40	26.0*
150	1:60	21.0*
150	1:80	19.0*
150	1:100	17.0*
150	1:150 (min. 5 WCs)	15.0 (at peak flow rate > 1 l/s)

Notes: * Denotes approximate values taken from Diagram 7 of the Approved Document, Section 2 of H1, for foul drains running 0.75 proportional depth. Minimum pipe diameters: 75 mm for waste water and 100 mm for soil water or trade effluent.

settlement may occur a suspended drainage system or other solution may be necessary.

- A pipe within 300 mm of the underside of a slab should be encased in concrete and made integral with the slab.
- Where pipes need to pass through a wall or foundation then a sleeve (minimum 50 mm clearance) or rocker pipes should be used to retain drain flexibility at these points.
- A drain trench within 1 m of a building should be backfilled with concrete up to the foundation soffit level. Where 1 m or more from the building the trench should be filled with concrete to within that dimension to the foundation soffit (less 150 mm).
- Consideration should also be given to the surcharging of drains or where rodent infestation may cause a problem. The local authority should be able to confirm the extent of any problem and precautions to be taken.
- Where a sewer connection by gravity is impracticable a pumped system may be used; see BS 8301 *Code of practice for building drainage.*
- A sewer connection should be made obliquely or in the direction of flow.

Bedding and backfilling

- Choice will be dependent on the pipe depth below ground level, the size and strength of the pipe, and the extent (or weight) of backfilling over the pipe.
- For **rigid pipes** of standard strength adopt 100 mm granular fill

bedding and 150 mm cover of selected backfill. For detailed provisions refer to Diagram 8 and Table 8 of the Approved Document. If pipes have less than the cover recommended in Table 8 then minimum 100 mm concrete encasement may be necessary, with movement joints at each pipe socket.

- For **flexible pipes** adopt 100 mm granular fill bedding and surround with 300 mm cover of selected backfill, as described in Diagram 9 of the Approved Document. The minimum depth of a drain under a road should be 900 mm (unless bridged or encased in reinforced concrete) and 600 mm under fields or gardens (unless bridged with paving slabs), maximum depth 10 m.

Blockage clearance

- The provision of access points to clear blockages assumes that traditional methods of rodding are used (which need not be in the direction of flow) and not mechanical or other means of clearing, which may be justifiable in certain circumstances.
- Access points are either: a rodding eye (capped extension of the drain); an access fitting (with no open channel), an inspection chamber (ground-level working space); or a manhole (with drain-level working space). The minimum dimensions are listed in Table 10.3, noting that due allowance should be made to accommodate all branch connections.
- Access points should be sited: on or near the head of each drain run; at a bend or change of gradient; at a change of pipe size; and at a

Table 10.3 Minimum dimensions for access points[a]

Type	Depth to (m)	Internal sizes (mm)	Spacing (m) to IC	MH
Rodding eye	–	Minimum 100	45	45
Small access fitting	0.6 or less	150 x 100 (150 dia)	22	22
Large access fitting	0.6 or less	225 x 100	45	45
Inspection chamber (IC)	0.6 or less	190 dia (max 150 drain)	45	45
	1.0 or less	450 x 450 (450 dia)	45	45
Manhole (MH)	1.5 or less	1200 x 750 (1050 dia)	45	90
[600 x 600 mm or over	over 1.5	1200 x 750 (1200 dia)	45	90
600 mm dia cover size]	over 2.7	1200 x 840 (1200 dia)	45	90
Shaft	over 2.7	900 x 840 (900 dia)	45	90

Notes: Covers to ICs and MHs should be removable and non-ventilating, of durable material (e.g. cast iron, steel or uPVC), and of suitable strength, bearing in mind location. MHs deeper than 1 m should have step irons or ladder to allow for access.

junction (unless clearance is possible from an access point, noting that rodding may only be possible from one direction). For the maximum spacing of access points on long drain runs reference should be made to Table 10 of the Approved Document, although spacings to inspection chambers and manholes have been included in Table 10.3.

A range of materials may be used for pipes and access points: clay, concrete or grey iron for rigid pipes; uPVC for flexibly jointed pipes; and brick, concrete or plastic for access points. Where possible, flexible joints should be adopted, which allow for any differential settlement of the pipes. Consideration may also need to be given to the separation of different metals to avoid electrolytic corrosion. Once laid, the drains and access points should exclude groundwater and rainwater and withstand a water test equal to a pressure head of 1.5 m above the drain invert (maximum 4 m to avoid damage to the drain). Alternatively an air test can be used to ensure a maximum loss of head on a manometer of 25 mm in 5 min for a 100 mm gauge (12 mm for a 50 mm gauge). Also refer to **Regulation 16** concerning the testing of drains. An **alternative approach** to the guidance above is to refer to BS 8301: 1985 *Code of practice for building drainage*: this includes a discharge unit method of determining pipe sizes.

REQUIREMENT H2: CESSPOOLS, SEPTIC TANKS AND SETTLEMENT TANKS

Any cesspool, septic tank or settlement tank shall be:

(a) of adequate capacity and so constructed that it is impermeable to liquids;
(b) adequately ventilated;(c) so sited and constructed that:
   (i) it is not prejudicial to the health of any person,
   (ii) it will not contaminate any underground water or water supply, and
   (iii) there are adequate means of access for emptying.

As an alternative to a public or private sewer connection, which may not be available, a cesspool, septic tank, settlement tank or small sewage treatment works may be used. Cesspools are basically large enclosed effluent storage tanks, which are emptied periodically. Whereas septic tanks and settlement tanks treat the sewage internally and discharge it to the subsoil via a system of land drains or a filter bed, desludging is normally required about once a year. The design and installation of a small sewage treatment system should comply

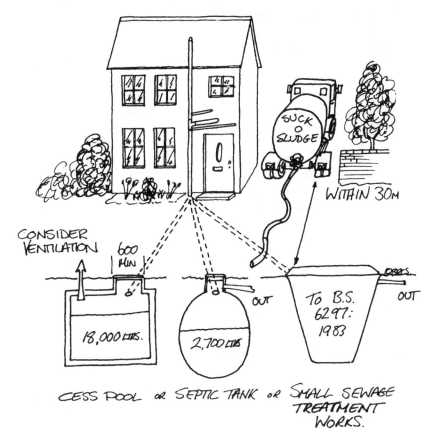

Figure 10.2   Alternatives for sewage disposal.

with the recommendations of BS 6297: 1983. The main provisions and alternatives are illustrated in Figure 10.2.

The design considerations for cesspools, septic tanks and settlement tanks can be summarized as follows.

- They should have sufficient capacity, below the inlet level, as indicated in Figure 10.2.
- They should be located to prevent the emptying, desludging and cleaning operations from causing a hazard to the building occupants or the removal of contents through a dwelling or work space. Access may be via an open covered space.
- They should prevent leakage of contents and the ingress of subsoil water.
- They should be constructed of brickwork (minimum 220 mm engineering bricks in 1:3 mortar), in-situ concrete (minimum 150 mm

thick of C/25/P mix), or glass-reinforced plastic, polyethylene or steel, where compliance with a current Agrément Certificate should be shown where possible.

- They should have suitable concrete covers and ventilation (septic tanks and settlement tanks may be open and fenced in).
- Where covered, access provision should be made for emptying, desludging and cleaning. Access covers should be of durable material (noting contents), minimum 600 mm dimension and be lockable.
- The inlet to a cesspool and both the inlet and outlet to a septic tank or settlement tank should have access provision for inspection.
- To minimize turbulance in a septic tank the inlet flow rate should be limited: for example, by the use of a dip pipe (for a tank up to 1200 mm in width) or a 1:50 fall for the last 12 m of the drain run (drains up to 150 mm).

The **alternative approach** to comply with Requirement H2 is to follow the relevant recommendations of BS 6297: 1983 *Code of practice for design and installation of small sewage treatment works and cesspools.* Formulae are included for the sizing of tanks.

REQUIREMENT H3: RAINWATER DRAINAGE

Any system which carries rainwater from the roof of the building to a sewer, a soakaway, a water course or some other suitable rainwater outfall shall be adequate.

The wording of the requirement is quite clear in that the surface water drainage system should:

- convey the roof water flow to a suitable outfall;
- minimize the risk of leakage and blockages;
- be made accessible for blockage clearance.

As with foul water drainage it is important to note that the requirement seeks to control the drainage system up to the outfall *only* and *not* the suitability of the outfall itself. Guidance from the local water authority may be necessary in this regard. For example, an existing surface water sewer may not be of a sufficient capacity to accept the proposed flow rates from the building; means to retain and restrict water outflow could therefore be required. Figure 10.3 serves to illustrate the provisions.

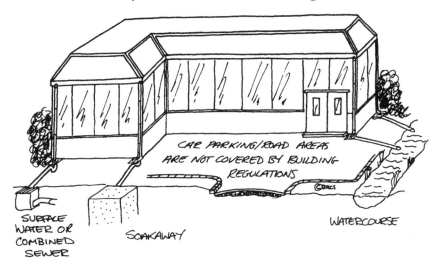

Figure 10.3 Roof drainage disposal provisions. The drainage of small individual roof areas of 6 m² or less is not controlled.

Section 1: Gutters and rainwater pipes

To calculate the effective roof area(s) to be drained the following formula can be used:

Effective design area (m^2) = plan area × multiplication factor (MF)

where the MF for a flat roof = 1, the MF for a pitched roof at 30° = 1.15, at 45° = 1.40, at 60° = 2.00, and at over 70° or any wall = 0.50.

Once the roof area to be drained has been established, the gutter and outlet (or downpipe) sizes can be found from Table 10.4. For more detailed design guidance reference should be made to BS 6367: 1983.

Table 10.4 Gutter and outlet sizes

Max effective roof area (m^2)	Gutter size (mm dia)	Outlet size (mm dia)	Flow capacity (l/s)
6	–	–	–
18	75	50	0.38
37	100	63	0.78
53	115	63	1.11
65	125	75	1.37
103	150	89	2.16

Note: Table relates to a half-round gutter, laid level, sharp-edged outlet at one end and where the distance to the outlet from a stop end does not exceed 50 times the water depth.

The remaining provisions are as follows.

- Gutters should be laid to allow overflow away from the building.
- Rainwater pipes should discharge to a drain or gulley, or to another gutter or other drained surface (taking care to avoid water flow over a pedestrian route).
- Discharge to a combined drain should be via a trap.
- A rainwater pipe serving more than one roof area should have the equivalent cross-sectional area of all the outlets.
- Gutters should remain watertight, and internal pipes should be able to withstand the air test as described for sanitary pipework.
- Gutters and downpipes should be of durable materials, with adequate strength and fixings to allow for thermal movement. Consideration may also need to be given to the separation of different metals to avoid electrolytic corrosion.

BS 6367: 1983 *Code of practice for drainage of roofs and paved areas* is referred to as the **alternative approach**.

Section 2: Rainwater drainage

As with foul water the below-ground surface water drainage system comprises the necessary pipes and fittings to connect the rainwater pipes to the outfall(s). Surface water drains may discharge to a combined public sewer that carries both foul and surface water. Where the public combined sewer (or private drain) cannot accommodate the additional flow a separate system with its own outfall should be installed. For all provisions relating to pipes, bedding and backfilling, blockage clearance, materials and testing direct reference should be made to the design guidance contained in Section 2 under Requirement H1. The following items are the exceptions and apply to drainage systems carrying rainwater only.

- The drainage pipes should have sufficient capacity to carry the anticipated flow from the roof and any runoff from paved or hard surfaces (taken at 50 mm per hour) even though these are not controlled under the regulations. Table 10.5 gives specimen values.
- The use of petrol interceptors may need to be considered for parking areas, although this is not controlled under the regulations.
- Where a gravity connection is impracticable or the sewer is liable to surcharge, water-lifting equipment will be necessary: see BS 8301: 1985.

As with foul water the **alternative approach** is to follow the relevant recommendations of BS 8301: 1985 *Code of practice for building drainage*: this includes additional detailed design and construction information.

Table 10.5 Minimum gradients and maximum capacities for surface water drains

Pipe diameter (mm)	Gradient	Flow rate (l/s)*
75	1:10	9.0
75	1:50	3.9
75	1:100 (minimum fall)	2.8
100	1:10	20.0
100	1:50	8.6
100	1:100 (minimum fall)	6.0
150	1:50	26.0
150	1:100	18.0
150	1:150	16.0

Notes: * Denotes approximate values taken from Diagram 1 of the Approved Document, Section 2 of H3, for rainwater drains running full. Minimum pipe diameter 75 mm. Capacity can be increased by increasing fall.

REQUIREMENT H4: SOLID WASTE STORAGE

1. Adequate means of storing solid waste shall be provided.
2. Adequate means of access shall be provided:

    (a) for people in the building to the place of storage, and
    (b) from the place of storage to a street.

The storage and collection of refuse to all buildings should not be prejudicial to health, should have sufficient capacity having regard to frequency of removal, and should be readily accessible from the street. The Approved Document guidance is split into two parts relating to domestic and non- domestic developments.

Domestic developments

Based on a refuse output of 0.09 m^3 per dwelling collected weekly, any house, flat or maisonette in a low-rise development up to four storeys should have, or have access to, a portable individual waste container able to hold at least 0.12 m^3 of refuse or a communal container with a capacity of at least 0.75–1 m^3. Floors above four storeys should be served by a refuse chute discharging to a communal container unless the siting or operation of a chute is impracticable. For example, site levels may make it difficult to locate a suitable chamber for the refuse containers, or a particular development may have the benefit of a 24 hour porter service. Further limited design guidance is given as follows.

- Individual containers should have close-fitting lids.
- Refuse chutes should have a non-absorbent internal surface and close-fitting access doors (located at each floor level containing a dwelling), and should be ventilated top and bottom. Cross-reference should also be made to Approved Documents B and E.
- Any room or chamber used for containers need not be enclosed, but if it is then high- and low-level ventilation should be provided, access allowed for filling and emptying (minimum 150 mm clearance around containers) and 2 m headroom where communal containers are located.
- The maximum travel distance for householders to carry refuse to a container (including a communal container) or chute should be 30 m.
- Refuse collection vehicle access should be within 25 m of the containers.
- Containers, including dustbins and communal containers, to new buildings only should be located so as to avoid collection through a building, unless it is a garage, carport or other covered open space. For conversions it may be necessary for the householder to bring the refuse and/or container to the front of the property for collection.
- Further detailed guidance with regard to refuse chute systems can be found in BS 5906: 1980.

Non-domestic developments

For other building types and where high refuse densities may occur, consultations should take place with the refuse-collecting authority, where proposals to address the following matters should be established:

- volume, nature and storage capacity required based on collection frequency and size/type of container;
- storage and on-site treatment related to layout and building density;
- location and access, for vehicles and operatives, to storage and treatment areas;
- hygiene arrangements and protection measures against fire hazards.

The **alternative approach** is to refer to BS 5906: 1980 *Code of practice for storage and on-site treatment of solid waste from buildings.*

CHAPTER 11

Approved Document J: Heat-producing appliances

The potential risks associated with the installation of heat-producing appliances in buildings are related to noxious fumes and fire. To combat these problems Requirements J1-J3 seek to provide for sufficient combustion air, the suitable discharge of combustion products and the protection of the building fabric.

The Approved Document guidance is split into four sections, starting with some basic provisions that apply generally. The remaining sections deal with solid fuel, gas and oil appliances up to certain specific ratings. Above these ratings specialist advice and/or the alternative approaches may be utilized. It should be noted that electrical heat-producing appliances and portable heaters are not controlled; in addition it is assumed that incinerators burning any fuel are controlled.

REQUIREMENT J1: AIR SUPPLY

Heat-producing appliances shall be so installed that there is an adequate supply of air to them for combustion and for the efficient working of any flue-pipe or chimney.

REQUIREMENT J2: DISCHARGE OF PRODUCTS OF COMBUSTION

Heat-producing appliances shall have adequate provision for the discharge of the products of combustion to the outside air.

REQUIREMENT J3: PROTECTION OF BUILDING

Heat-producing appliances and flue-pipes shall be so installed, and fireplaces and chimneys shall be so constructed, as to reduce to a reasonable level the risk of the building catching fire in consequence of their use. The requirements in this Part apply only to fixed heat producing appliances which:

(a) are designed to burn solid fuel, oil or gas; or

(b) are incinerators.

Section 1: Provisions which apply generally

Irrespective of the appliance rating, consideration should be given to the following items.

- An appliance should be **room-sealed** (i.e. not reliant on combustion air from within the room), or the room or space in which it is located should be provided with a ventilation opening. This may be via an adjoining room or space that also has the same-size opening giving to the external air.
- Normally ventilation openings should not be in fire-resisting walls; this may cause problems where the use of a fire shutter and/or fuel cut-off could be used.
- For appliances that are not room-sealed, reference should be made to Approved Document F concerning the installation of air extract fans.
- An appliance should have a balanced flue or low-level flue, or should connect with a flue pipe or chimney that discharges to the external air, unless the appliance can operate without discharging the products of combustion to the outside.
- A flue may only have an opening into it for inspection and cleaning, and the fitting of an explosion door, draught stabilizer or diverter.
- A flue may serve more than one appliance in a room but should not serve appliances or have openings (except for inspection and cleaning) in any other rooms.
- A chimney built prior to 1 February 1966 may be used subject to its showing no signs of being unsuitable.

Section 2: Additional provisions for solid fuel-burning appliances with a rated output up to 45 kW

Solid fuel appliances are basically open fires, which utilize an open

chimney flue, or closed appliances, e.g. boilers, which should use a
flue pipe to connect the appliance to a chimney. The guidance
contained in the Approved Document can be summarized as follows.

Air supply and flues

- Air supply, flue sizes and flue outlet locations should be in accordance with Figure 11.1.
- With the exception of a horizontal connection to a chimney, up to
  150 mm, flues should be vertical. Where a bend is necessary it
  should not be at an angle of more than 30° to the vertical. Where
  an offset is necessary the flue size should be increased by 25 mm in
  each direction.
- A flue pipe should not pass through a roof void.
- Flue pipes may be of cast iron, 3 mm mild steel, 1 mm stainless steel,
  or vitreous enamelled steel. Spigot and socket joints, where used,
  should be fitted with sockets upwards.

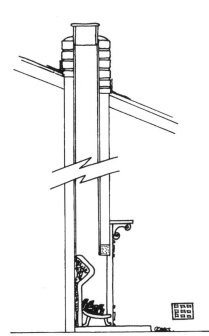

Minimum size of chimney flue for:

Fireplace recess, < 500 × 550 mm opening	200 mm dia. or equivalent
Fireplace recess, > 500 × 550 mm opening	15% of recess opening
Closed appliance up to 20 kW rated output	125 mm dia or equivalent
Closed appliance up to 20 kW burning bituminous coal or 20–30 kW	150 mm dia or equivalent
Closed appliance 30–45 kW	175 mm dia or equivalent

For flue pipes and chimneys never less
than the appliance flue outlet.

Locate flue outlet:
1 m above roof or roof openings, 600 mm
at ridge of pitched roof and adjacent to or
adjoining buildings

Air supply:
Open appliance: minimum 50% of throat
opening area (BS 8303: 1980)

Other appliances: 550 mm²/kW of rated
output above 5 kW (add 300 mm²/kW
where a flue stabilizer is fitted)

Figure 11.1   Appliances burning solid fuel (up to 45 kW output rating).

- Flue pipes should be isolated from the building fabric by an air space of at least 3 x flue diameter or 200 mm of solid non-combustible material, or by using a non-combustible shield, 3 x diameter of flue in width, where the combustible material is isolated from the shield by 12.5 mm and the flue by 1.5 x diameter.

Chimneys

- Should retain structural stability and performance at 1100°C.
- Should be provided with a debris collection space where the chimney is not directly above the appliance.
- Brick chimneys should be lined with clay flue liners, imperforate clay pipes or high alumina cement and kiln-burnt or pumice aggregate pipes. It is important to use fireproof mortar and to fill the space between the liner and brickwork with weak mortar or insulating concrete.
- Blockwork chimneys should be lined as for brick chimneys or made of refractory material, high-alumina cement/kiln-burnt or pumice aggregate.
- Minimum wall thickness for both brickwork and blockwork is 100 mm, or 200 mm where the wall is between the flue and another compartment, building or dwelling.
- Isolate combustible materials by at least 200 mm from the flue or 40 mm from the chimney face itself (excluding floorboards, skirtings, etc.).
- Use can be made of **factory-made insulated chimneys**, which should accord with BS 4543: Parts 1 and 2: 1990 and BS 6461: Part 2: 1984, but should not pass through other buildings or enclosed spaces unless encased or guarded by non-combustible construction.

Location of appliances

- The constructional hearth on which the appliance is located should be of non-combustible material at least 125 mm thick, 840 mm square (for a freestanding appliance) or should project 500 mm in front of a fireplace recess.
- Fireplace recesses should be of solid non-combustible material.
- For specific design guidance on fireplace recesses, walls adjacent to hearths and location of appliances reference should be made to Diagrams 3–8 of the Approved Document.

The **alternative approach** to the above guidance is to refer to the relevant recommendations of BS 8303: 1986 *Code of practice for installation of domestic heating and cooking appliances burning solid mineral fuels.*

Section 3: Additional provisions for individually flued (non-fan-assisted) gas-burning appliances with a rated input up to 60 kW (and air supply for cooking appliances)

This section of the Approved Document offers guidance on open-flued and balanced-flued appliances, solid fuel effect fires and cookers all burning gas and up to a rated input of 60 kW. This guidance is described as follows.

Air supply and flues

- Air supply, flue sizes and flue outlet locations should be in accordance with Figure 11.2.
- Flues should be vertical and avoid horizontal runs. Where a bend is

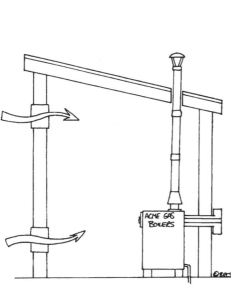

Minimum size of flue:

Gas fire: Minimum cross-sectional area of 12 000 mm² for a round flue, 16 500 mm² for a square flue, 90 mm minimum dimension

Other appliances: minimum cross-sectional area as appliance outlet

Locate flue outlet:

Balanced-flued appliance: to allow free intake of combustion air, dispersal of products of combustion, prevent entry of matter that could block flue, guard terminal if liable to damage or person contact and locate minimum 300 mm from any opening above terminal

Other appliances: located at roof level to allow air flow, minimum 600 mm from any opening and fitted with a flue terminal where dimension less than 175 mm (except a gas fire)

Air supply: Open-flued appliance: permanent vent opening minimum 450 mm²/kW of input rating exceeding 7 kW

Cooker: openable window, plus a permanent vent of 5000 mm² if room less than 10 m³

Figure 11.2 Appliances burning gas (up to 60 kW input rating). Open-flued appliances must not be used in a garage, bathroom or shower room.

necessary it should not be at an angle of more than 45° to the vertical.

- Installation of simulated coal or wood-effect gas fires should follow manufacturers' instructions or accord to BS 5871: Parts 1, 2 or 3: 1991.
- Flue pipes may be of cast iron, 3 mm mild steel, 1 mm stainless steel, vitreous enamelled steel, other sheet metals (as described in BS 715: 1986) or asbestos cement. Spigot and socket joints, where used, should be fitted with sockets upwards.
- Flue pipes should be isolated from combustible material by a minimum of 25 mm, or where passing through a wall, floor or roof, be isolated by a non-combustible sleeve with 25 mm air gap (measured from the inner pipe for a double-walled flue). Cross-reference should be made to Approved Document B where flues pass through compartment walls.
- Brick chimneys should be lined with clay flue liners, imperforate clay pipes or high-alumina cement and kiln-burnt or pumice aggregate pipes.
- Chimneys constructed with flue blocks should accord with BS 1289: Part 1: 1986 and Part 2: 1989.
- Minimum wall thickness for both brickwork and blocks is 25 mm.
- A flexible flue liner may be used in a chimney if it accords with BS 715: 1989 or the chimney was built before 1 February 1966 or is already suitably lined and constructed in accordance with the Approved Document.
- Use can be made of **factory-made insulated chimneys**, which should accord with BS 4543: Parts 1 and 2: 1990 and BS 6461: Part 2: 1984, but should not pass through other buildings or enclosed spaces unless encased or guarded by non-combustible construction.
- Chimneys should be provided with a debris collection space where the chimney is not lined or constructed of flue blocks.

Location of appliances

- The constructional hearth for a back boiler should be non-combustible material at least 125 mm thick, or 25 mm thick on 25 mm high supports. Note that a hearth is not needed where the flame of any appliance is above 225 mm or if the appliance complies with BS 5258 or BS 5386.
- For specific design guidance concerning hearths and appliance locations reference should be made to Diagrams 9 and 10 of the Approved Document.

The **alternative approach** is to refer to the relevant recommendations of one of the following British Standards: BS 5440: Part 1: 1990 and Part

2: 1989, BS 5546: 1990, BS 5864: 1989, BS 5871: Parts 1-3: 1991, BS 6172: 1990, BS 6173: 1990, BS 6798: 1987. British Gas also publish a range of guidance including, *Gas in Housing– a Technical Guide* (1990).

Section 4: Additional provisions for oil-burning appliances with a rated output up to 45 kW

Appliances that burn oil operate at varying temperatures, and the guidance that follows reflects this point.

Air supply, flues and chimneys

- Air supply, flue sizes and flue outlet locations should be in accordance with Figure 11.3.
- Flues should be vertical and avoid horizontal runs. Where necessary, a bend should not be at an angle of more than 45° to the vertical.
- Where flue gas temperatures, under worst operating conditions, could exceed 260°C then the provisions for flues and chimneys in

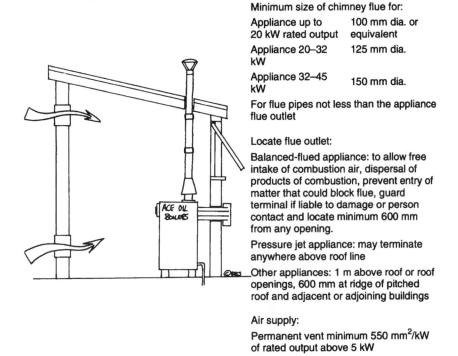

Minimum size of chimney flue for:

Appliance up to 20 kW rated output	100 mm dia. or equivalent
Appliance 20–32 kW	125 mm dia.
Appliance 32–45 kW	150 mm dia.

For flue pipes not less than the appliance flue outlet

Locate flue outlet:

Balanced-flued appliance: to allow free intake of combustion air, dispersal of products of combustion, prevent entry of matter that could block flue, guard terminal if liable to damage or person contact and locate minimum 600 mm from any opening.

Pressure jet appliance: may terminate anywhere above roof line

Other appliances: 1 m above roof or roof openings, 600 mm at ridge of pitched roof and adjacent or adjoining buildings

Air supply:

Permanent vent minimum 550 mm²/kW of rated output above 5 kW

Figure 11.3. Appliances burning oil (up to 45 kW output rating).

Section 2 should be applied. If this temperature is unlikely to be exceeded then follow the provisions outlined within Section 3.

- Use can be made of **factory-made insulated chimneys**, which should accord with BS 4543: Parts 1 and 2: 1990 and BS 6461: Part 2: 1984, but should not pass through other buildings or enclosed spaces unless encased or guarded by non-combustible construction.

Location of appliances

- Where the floor surface temperature under the appliance could exceed 100°C then a constructional hearth in accordance with Section 2 should be provided. Where this temperature is unlikely to be exceeded then the appliance may stand on a rigid and imperforate sheet of non-combustible material.
- The shielding of the appliance itself should accord with the guidance in Section 3 if the surface temperature of the back or sides of the appliance could exceed 100°C.

A British Standard is referred to as the **alternative approach**, namely BS 5410: Code of practice for oil firing: Part 1: 1977 Installations up to 44 kW output for space heating and hot water supply purposes.

The Approved Document concludes with a list of over 20 British Standards, which are referred to in respect of Requirements J1, J2 and J3.

CHAPTER 12

Approved Document K: Stairs, ramps and guards

People, including children and the elderly, should be able to use stairs safely and be protected from the risk of falling from one floor level to others below. The degree of safety provisions necessary depends on the purpose group of the building, the number of users, their familiarity with the building and the extent of access made available to the building. The guidance contained in Approved Document K reflects its close relationship with Approved Document B, relating to fire safety and means of escape, and Approved Document M, covering access to buildings for the disabled. An example of this is the omission of minimum stair and ramp widths, which are given in Approved Documents B and M.

The Workplace (Health, Safety and Welfare) Regulations 1992 – main requirements, relating to building design, are now covered by the Building Regulations. Compliance with Requirements K1 (and M2), K2, K3, K4 and K5, in accordance with Section 23(3) of the Health and Safety at Work, etc. Act 1974, would prevent the service of an improvement notice under the Workplace Regulations.

For mixed use developments (including dwellings) the requirements for non-domestic use can be applied to the shared parts of the building.

REQUIREMENT K1: STAIRS, LADDERS AND RAMPS

Stairs, ladders and ramps shall be so designed, constructed and installed as to be safe for people moving between different levels in or about the building.
Requirement K1 applies to stairs, ladders and ramps which form part of the building.

The wording of the requirement and the limitations of application, as illustrated in Figure 12.1, are relatively clear. Differences in level less

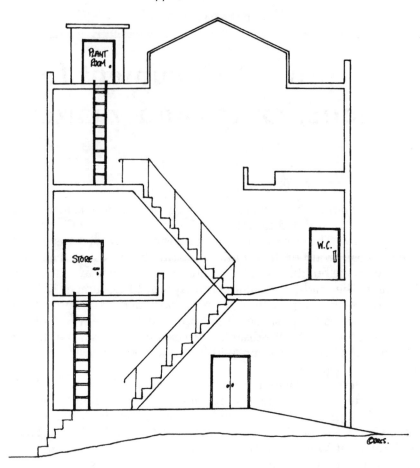

Figure 12.1 Application of stair and ramp provisions.

than 600 mm in a dwelling, and up to two risers (or 380 mm, if not part of a stair) in other buildings are not controlled. Also, steps or ramps on land surrounding a building are not controlled, although an entrance ramp or flight of steps would be controlled, as they can be regarded as part of the building. Section 1 of the Approved Document deals with the design of stairs, and ramps are covered in Section 2. With specific regard to assembly buildings, containing gangways serving spectator areas, reference should also be made to the following guidance sources:

- BS 5588: Part 6: 1991 *Code of practice for places of assembly.*
- Home Office (1990) *Guide to Fire Precautions in Existing Places of Entertainment and Like Premises.*
- Home Office (1997) *Guide to Safety at Sports Grounds.*

A series of definitions in relation to the Requirements are listed:

Flight – the part of the stair or ramp between landings formed by a continuous series of steps or a continuous ramp.
Helical stair – a helix round a central void.
Pitch line – notionally connects all tread nosings of a flight.
Spiral stair – a helix round a central column.
Stair – a succession of steps and landings that allow access by foot to other levels.

Section 1: Stairs

The first design criterion to consider is the steepness of stairs. Three categories of stairs are specified with corresponding maximum rise and minimum going limitations (the rise is measured between the top surfaces of each tread, and the going dimension is measured from nosing to nosing, all risers and goings in a flight should be the same):

- **private stair**, serving only one dwelling, maximum 220 mm rise and minimum 220 mm going, and a maximum pitch of 42°;
- **institutional and assembly stair**, serving places where a substantial number of people will gather, maximum 180 mm rise and minimum 280 mm going (250 mm going for a floor area not exceeding 100 m^2);
- **other stair**, serving any other building, maximum 190 mm rise and minimum 250 mm going.

The relationship, $2 \times$ rise + going = 550 mm − 700 mm can also be utilized for design purposes. For gangways serving seated spectators the maximum pitch should not exceed 35°. An **alternative approach** is also stated with regard to steepness, namely BS 5395 *Stairs, ladders and walkways* Part 1: 1977 *Code of practice for the design of straight stairs.*

The other provisions contained in Section 1 can be summarized as follows.

- Open riser stairs should have a minimum tread overlap of 16 mm, and where these stairs are likely to be used by children under 5 years no gap should allow the passage of a 100 mm diameter sphere (the approximate size of a child's head).
- Access routes between levels, including flights of stairs, should have a minimum headroom of 2 m. For a stair serving a loft conversion this may be reduced to 1.9 m at the centre of the stair (1.8 m to side of stair).
- For minimum stair widths reference should be made to Approved Documents B and M, noting that these do not offer guidance on the minimum stair widths *within* dwellings.
- The width of a flight, in a public building, exceeding 1800 mm

should be subdivided so that each flight does not exceed this dimension.

- Stair flights should be limited to 16 risers where they serve areas utilized for shop or assembly purposes. For other stairs that incorporate a flight with more than 36 risers a change of direction of at least 30° should be introduced by way of a landing, the dimensions of which should be at least the width of the stair.
- A clear, level and unobstructed landing should be provided at both the top and bottom of each flight, the dimensions of which should be at least the width of the stair. Two concessions to this are allowable, in that any door may swing across a landing located at the **bottom of the flight** and any cupboard or duct doors may swing across a landing at the top or bottom of the flight. For both concessions a minimum 400 mm landing depth, to the full width of the stair, should be retained.
- Landings formed by the ground may be at a gradient not steeper than 1 in 20. Since differences in level less than 380 mm (600 mm for dwellings) are not controlled, landings at external door locations, for example, need not be provided. Consideration should, however, be given to the safe use of the door, with special regard to satisfactory means of escape.

Section 1 isolates particular provisions for special stairs, which can be listed as follows.

- The going for steps with tapered treads should be measured at the centre, for stair widths not exceeding 1000 mm, or 270 mm in from each side, for a width 1000 mm or over, minimum tread width 50 mm.
- Consecutive tapered treads should all have the same going. Tapered treads combined with a straight flight should have the same going as the straight flight.
- Tapered tread stairs can also be designed to BS 585: Part 1: 1989.
- For spiral and helical stairs reference should be made to BS 5395: Part 2: 1984 *Code of practice for the design of helical and spiral stairs*. The guidance for the steepness of the stairs will also apply.
- Reduced goings, for spiral and helical stairs, may be acceptable in conversion work where space is limited and the stair only serves one habitable room.
- An **alternating tread stair** may only be used in a loft conversion to give access to one habitable room, with a bathroom and/or WC (but not the only WC). Space is saved by alternate cutaway treads which retain the required rise and going dimensions. Treads should have slip-resistant surfaces, and handrails are necessary to both sides of the stair.
- Where no other option is available, consideration can be given to the

use of a fixed ladder to serve just one habitable room. It should have handrails both sides and not be retractable if used for means of escape.
- For industrial buildings specific design guidance may be necessary, where reference should then be made to BS 5395: Part 3: 1985 *Code of practice for the design of industrial stairs, permanent ladders and walkways*, or BS 4211: 1987 *Specification for ladders for permanent access to chimneys, other high structures, silos and bins.*

Guidance is given on handrails and guarding to stairs:

- For a stair width not exceeding 1000 mm a handrail should be provided to one side, and to both sides for a width 1000 mm or over. Handrails need not be provided to the two bottom steps unless the stair is in a public building or it could be used by disabled people.
- Handrail heights should be between 900 mm and 1000 mm to all buildings, measured from the pitch line or floor up to top of handrail.
- Stairs, including flights and landings, should be guarded at the sides as outlined in Table 12.1.

Table 12.1 Guarding design

Building category	*For the location specified use guarding height of:*		
	800 mm	*900 mm*	*1100 mm*
Single-family dwellings		Stairs, landings, ramps, edges of internal floors	External balconies and roof edges
Factories and warehouses Residential, office, institutional, retail, educational and public buildings		Stairs and ramps For flights	Landings and floor edges All other locations
Assembly	530 mm in front of fixed seating	For flights	All other locations
All buildings (except roof windows in loft conversions)	Opening windows and glazing at changes of level		

- For stairs likely to be used by children under 5 years the guarding should have no gap that would allow the passage of a 100 mm diameter sphere; a maximum 50 mm gap between the lower edge of the guarding and the pitch line would be acceptable for this location. Children should also not be able to climb the guarding.
- Finally in Section 1, less demanding provisions apply for access to areas of maintenance where the frequency of visits is taken into account. Access at least once a month could utilize domestic stairs or where less frequent access is required a portable ladder could be used (covered by the Construction (Design and management) Regulations 1994).

Section 2: Ramps

Since ramps are often used to allow disabled people access and circulation within the building, close cross-reference must be made to Approved Document M, Sections 1 and 2. For the design of ramps cross-reference should also be made to Section 1, with specific regard to landings and guarding. The remaining provisions particular to ramps can be listed as follows:

- The steepness of a ramp should not exceed 1 in 12 to allow its safe use by all people, including those disabled and using a wheelchair.
- Throughout the length of the ramps and landings a minimum headroom of 2 m should be provided, and ramps themselves should be kept clear of permanent obstructions.
- Landings are needed, but guidance on their spacing, at the top and bottom of ramp flights, is not provided in the Approved Document.
- Handrails for ramps, with a rise greater than 600 mm, should give adequate support; should allow a firm grip; should be positioned at a height of between 900 mm and 1000 mm; and should be provided to one side, or to both sides if width is 1000 mm or over.

REQUIREMENT K2: PROTECTION FROM FALLING

(a) Any stairs, ramps, floors and balconies and any roof to which people have access, and
(b) any light well, basement area or similar sunken area connected to a building,
shall be provided with barriers where it is necessary to protect people in or about the building from falling.

Requirement K2(a) applies only to stairs and ramps which form part of the building.

REQUIREMENT K3: VEHICLE BARRIERS AND LOADING BAYS

1. Vehicle ramps and any levels in a building to which vehicles have access, shall be provided with barriers where it is necessary to protect people in or about the building.
2. Vehicle loading bays shall be constructed in such a way, or be provided with such features, as may be necessary to protect people in them from collision with vehicles.

Section 3 deals with the requirements for both pedestrian guarding and vehicle barriers, the application of which is illustrated in Figure 12.2.

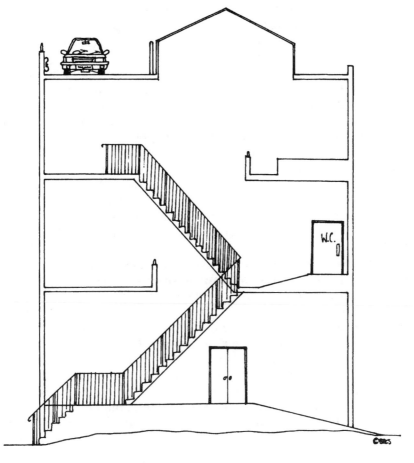

Figure 12.2.   Application of guard and barrier provisions.

Section 3: Guards and barriers

The first factor to consider for pedestrian guarding is where it should reasonably be provided so as to prevent people from being injured by falling from a height above 600 mm in a dwelling, and from falling in the first place in any other building above a height of two risers (or 380 mm, if not part of a stair). The locations that should be considered are: the edge of any floor (including an opening window), gallery, balcony, roof (including rooflights and other openings), places to which people have access (unless for the purpose of maintenance or repair only), any light well, basement or similar sunken area next to or within a building. Guarding need not be provided on ramps used only for vehicle access or to loading bays where it would present an obstruction to normal use of the area.

The design of the guarding may take the form of a wall, parapet, balustrade, screen or similar obstruction, which should be at a minimum height as indicated in Table 12.1. From a structural point of view the guarding should resist a horizontal force applied at the top of the guarding: reference should be made to BS 6399: Part 1: 1996, and BS 6180: 1995 for infill panels.

Where the guarding utilizes glazing, reference should be made to Approved Document N. Finally, in the design of pedestrian guarding, consideration should be given to whether the building is likely to be used by children under 5 years. If so, no gap in the guarding should allow the passage of a 100 mm diameter sphere, and children should not be able to climb the guarding.

Less demanding provisions apply for access to areas of maintenance where the frequency of visits is taken into account. Access at least once a month could utilize domestic stairs or where less frequent access is required temporary guarding and/or warning notices could be used (covered by the Construction (Design and management) Regulations 1994).

Turning to the need for vehicle barriers, these should be provided to the edges of any floor, roof or ramp to which vehicles have access at or above ground or upper floor levels. Barrier design may take the form of a wall, parapet, balustrade, screen or similar obstruction, which should be at a minimum height, for any building, of **375 mm to a floor or roof edge** and **610 mm at a ramp edge**. The barrier should resist a horizontal force applied at the top of the barrier, where reference should be made to BS 6399: Part 1: 1996.

Loading bays should have at least one exit point, i.e. steps, or two exit points for wide loading bays used by more than three vehicles. A refugee could also be used as an alternative, to avoid people being crushed.

REQUIREMENT K4: PROTECTION FROM COLLISION WITH OPEN WINDOWS, ETC.

Provision shall be made to prevent people moving in or about the building from colliding with open windows, skylights or ventilators.
Requirement K4 does not apply to dwellings.

Section 4: Protection from collision

The requirement basically applies the guidance contained in Approved Document M, although it does not apply to dwellings.

Options to protect the projecting parts of windows, skylights and ventilators more than about 100 mm, either internally or externally, are described:

- Keep projection at least 2 m above ground or floor level; or
- If below 2 m, provide a feature, barrier or rail, about 1100 mm high; or
- By using tactile differences, cobbles or a planting strip.
- For spaces used only infrequently and for maintenance, reliance could be placed on clear marking of the projection.

REQUIREMENT K5: PROTECTION AGAINST IMPACT FROM AND TRAPPING BY DOORS

1. Provision shall be made to prevent any door or gate:

    (a) which slides or opens upwards, from falling onto any person: and
    (b) which is powered, from trapping any person.

2. Provision shall be made for powered doors and gates to be opened in the event of a power failure.
3. Provision shall be made to ensure a clear view of the space on either side of a swing door or gate.

Requirement K5 does not apply to:
(a) dwellings, or
(b) any door or gate which is part of a lift.

Section 5: Protection against impact

A series of options to prevent the opening and closing of doors and gates presenting a safety hazard are described:

- Provide a visibility zone of glazing to cover a height of between 900 mm and 1500 mm above the floor level, unless low enough to see over, i.e. 900 mm;
- Provide effective stops at ends of sliding tracks and retaining rail should suspension system fail;
- Provide effective device to prevent an upward opening door or gate (e.g. roller shutter) from falling;
- Provide safety features to power operated doors and gates, i.e. pressure sensitive door edge/stop switch/opening override in case of power failure.

CHAPTER 13

Approved Document L: Conservation of fuel and power

The United Nations Framework Convention on Climate Change was ratified by the United Kingdom in December 1993. This will require the UK to achieve 1990 levels of greenhouse gas emissions by the year 2000. As part of the UK programme to reduce carbon dioxide emissions and to encourage the efficient use of energy the requirements of the Building Regulations have been extended and the guidance contained in Approved Document L revised. This takes account of the fact that almost half of the total carbon dioxide emission levels in the UK are associated with the energy used in buildings, 60% of which relates to energy used in the home.

The changes to the regulations themselves are brought into effect by the Building Regulations 1994 (Amendment) Regulations and are discussed in Chapter 1. Requirement L1 is now applied to a material change of use as well as all new building work. In addition, the new Building Regulation 16 seeks the provision of an energy rating for each new dwelling created, either by new build or change of use.

The wording, application and limits of application for Requirement L1 are relatively straightforward, although there are a number of instances where the degree of application needs to be considered.

- **Small extensions to dwellings**, which do not exceed 10 m^2 in floor area, can adopt the construction of the existing dwelling.
- Buildings that do not need heating or have only a **low level of heating** (system output does not exceed 50 W/m^2 for industrial and storage buildings, 25 W/m^2 for other buildings excluding dwellings) would not require insulation to the building fabric. The building should be insulated and sealed if the heating level is not known.
- **Large complex buildings** may benefit from assessing each part separately, although this may not always be possible: for example, a large open atrium space may contain a range of separate uncompartmented uses.

REQUIREMENT L1: CONSERVATION OF FUEL AND POWER

Reasonable provision shall be made for the conservation of fuel and power in buildings by:

(a) limiting the heat loss through the fabric of the building;
(b) controlling the operation of the space heating and hot water systems;
(c) limiting the heat loss from hot water vessels and hot water service pipework;
(d) limiting the heat loss from hot water pipes and hot air ducts used for space heating;
(e) installing in buildings artificial lighting systems which are designed and constructed to use no more fuel and power than is reasonable in the circumstances and making reasonable provision for controlling such systems.

Requirements L1 (a), (b), (c) and (d) apply only to:

(a) dwellings;
(b) other buildings whose floor area exceeds 30 m^2.

Requirement L1 (e) applies only within buildings where more than 100 m^2 of floor area is to be provided with artificial lighting and does not apply within dwellings.

The technical guidance of the Approved Document is split into two sections dealing with dwellings (including flats) and buildings other than dwellings, concluding with a series of eight appendices. Before moving on to the contents of these sections a number of terms need to be defined.

Thermal conductivity or **λ value**, is the rate at which heat will pass through a material, and is expressed in watts per metre per degree of temperature difference (W/mK).

Thermal transmittance, or **U-value** is the rate at which heat will pass through a square metre of structure (or fabric) when the air temperatures on either side differ by one degree, expressed in watts per square metre per degree of temperature difference (W/m^{2}K). The calculation of U-values should take account of the effects of thermal bridging: e.g. timber joists, mortar joints, window frames, etc. Reference should be made to Appendix B of the Approved Document.

Areas of walls, floors and roofs should be taken from internal finished

dimensions; floor areas should take account of ducts, stairwells, etc.; and roofs should be measured at the plane of the insulation.

Exposed element an element of the building fabric that is exposed to the external air, and includes a ground bearing floor and a suspended floor, whether ventilated or not.

Semi-exposed element an element between a heated space and an unheated space with exposed elements. Examples include: the walls and/or floors separating a domestic integral garage from the remainder of the house; flats from an unheated stairway; or a heated office from an unheated store room or service duct.

Energy rating of dwellings using an approved procedure, for example the **standard assessment procedure (SAP)**, is required under Regulation 16 for each new dwelling created, either by new build or change of use. There is, however, no obligation to achieve a particular SAP rating, although ratings of between 80 and 85, subject to dwelling size, can be used to show direct compliance with Part L. On the other hand a low calculated rating (60 or less) will require the adoption of more stringent U-values if the elemental method of compliance is being utilized. For full details of the SAP calculation procedure reference should be made to Appendix G of the Approved Document.

Calculations for the SAP rating can be submitted to the building control authority by one of the following:

- the **applicant**, where building control would check the calculations;

- a **competent person**, who can certify calculations that may be acceptable subject to the question of competence being previously agreed with building control, who would remain responsible for enforcement;
- **assessors**, authorized by the Secretary of State, where calculations prepared by them as competent persons can be accepted by building control;
- an **approved person**, approved by the Secretary of State, who can certify compliance with Part L as a whole, and calculations prepared by them, including SAP ratings, *will* be accepted by building control.

Going hand in hand with energy conservation measures are a number of potential technical risks. Examples include parts of construction that remain colder, thus encouraging interstitial condensation, or different forms of construction which may lead to rain penetration. Guidance on how to avoid these associated risks can be found in the following publications, which are widely referred to in the Approved Document, but none of which have the status of approved documents:

- BRE Report 262 Thermal insulation: avoiding risks 1994 (Second Edition);
- NHBC Thermal Insulation and Ventilation Guide 1991;
- Approved Document F: Ventilation.

Section 1: Dwellings

To demonstrate compliance with Requirement L1, one of three methods, highlighted in the Approved Document, may be used (note that for new dwellings an SAP energy rating is required in any case):

- elemental method;
- target *U*-value method;
- energy rating method.

The **elemental method** can be regarded as the most straightforward, whereby the elements of construction, namely walls, floors, roofs, windows, doors and rooflights, should meet the specified thermal performance criteria, i.e. standard *U*-values. These values depend on whether the SAP rating of the dwelling is above or below 60. A summary of the provisions, including allowances for windows, doors and rooflights, is illustrated in Figure 13.1.

As a minimum starting point for design purposes it is suggested that the values for an SAP rating of over 60 are used. The actual SAP energy rating for the dwelling can then be established, from the guidance contained in Appendix G, and checked to see if it is actually

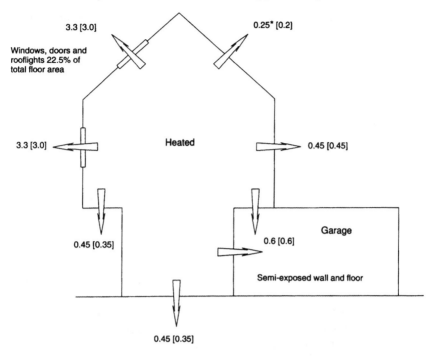

Figure 13.1 Standard U-values (W/m²K) and areas of openings for dwellings. *Notes:* U-values indicated are for dwellings with an SAP rating of over 60; values in brackets [ ] are for SAP ratings of 60 or less.
A U-value of 0.35 applies to a roof where there is no loft, as indicated by *. A roof at a pitch of 70° or over may have the same U-value as a wall.

over 60 as assumed. If not, then the values of the elements of construction should be modified so as to produce an SAP rating of over 60. To achieve the *U*-values stated in Figure 13.1, reference can be made to certified manufacturers' data or Appendix A of the Approved Document. The Appendix gives tables and example calculations to establish thicknesses of various insulation materials in different typical elements of construction. Examples of the procedures used for this method, including the calculation of *U*-values of ground floors, are contained in Appendices C and E of the Approved Document.

The **basic allowance** for windows, doors and rooflights, as stated in Figure 13.1, is seeking the provision of double glazing or secondary glazing to achieve an average *U*-value of 3.3 W/m²K. Table 2 of the Approved Document gives indicative *U*-values for various window, door and rooflight constructions depending on frame type, glazing air gap, glass type used, etc. Where enclosed, unheated and draughtproof porches or conservatories are used to protect windows and doors with

Table 13.1 Permitted variation in the area of windows and doors for dwellings

Average U-value (W/m²K)	Maximum permitted area as a % of total floor area, for SAP ratings of:	
	60 or less	Over 60
2.0	37.0	41.5
2.5	28.0	31.5
3.0	22.5	25.0
3.3	20.0	22.5
3.5	19.0	21.0
4.0	16.0	18.0
4.2	15.5	17.0

Note: For intermediate values reference should be made to Table 3 of the Approved Document.

single-glazed panels a U-value of 3.3 W/m²K can be assumed for these openings.

A **modification** to the basic allowance of 22.5% can be made to allow larger areas of windows and doors but not rooflights. This is subject to improvements being made to the average U-value of the openings in question. Table 13.1 outlines the permitted range of variation.

To show compliance with Requirement L1 for **domestic extensions** (exceeding 10 m²) the U-values appropriate to an SAP rating of over 60, as stated in Figure 13.1, should be applied. This would include the 3.3 W/m²K U-value for windows, doors and rooflights where the 22.5% allowance can apply to the floor area of the extension itself or the total floor area of the existing dwelling and the extension.

The **target U-value method** offers equations to calculate a target U-value of the exposed fabric as a whole, which is then compared with the average U-value of the actual dwelling. Semi-exposed elements should comply with the values in Figure 13.1, and are not incorporated within the equations. The target is first calculated using one of two equations dependent on the SAP rating:

Target U−value total (SAP 60 or less)

$$= \frac{\text{total floor area} \times 0.57}{\text{total area of exposed elements}} + 0.36$$

or

Target U–value total (SAP over 60)

$$= \frac{\text{total floor area} \times 0.64}{\text{total area of exposed elements}} + 0.4$$

The rate of heat loss per degree (area × U-value) is then calculated for each exposed element and the total put into the following equation to establish the average U-value:

$$\text{Average } U\text{–value} = \frac{\text{total rate of heat loss per degree}}{\text{total external surface area}}$$

where the **average U-value** < **target U-value** to show compliance.

The target U-value method assumes equal distribution of glazing to the north and south elevations and either a gas- or oil-fired hot water central heating system with a seasonal efficiency of at least 72%. Two options are given that can take account of improvements in these two areas (full details and examples of the procedures are given in Appendix F of the Approved Document):

- **Account for solar gains** can be made where areas of glazing on the south elevation exceed those on the north. The total window area can be taken as the actual window area less 40% of the difference in area of the glazing facing south and that facing north, ±30°.
- **Account for higher-efficiency heating systems** can be made where, for example, the installation of a condensing boiler would increase seasonal efficiency to 85%, allowing the target U-value to be increased by 10%.

The **energy rating method** takes account of ventilation rate, fabric losses, water heating requirements, internal heat gains and solar gains. To show compliance with Requirement L1 each dwelling, including separate flats in a block or conversion, should have an **SAP energy rating** as stated in Table 13.2. However, when using either of the calculation procedures the minimum U-values of exposed walls and exposed floors should not be allowed to be worse than 0.7 W/m²K and 0.35 W/m²K for roofs.

Table 13.2   Minimum SAP energy ratings for dwellings

Dwelling floor area (m²)	*SAP energy rating*
80 or less	80
80–90	81
90–100	82
100–110	83
110–120	84
Over 120	85

Also, to avoid additional heat losses and local condensation problems the detailing at **thermal bridging locations**, i.e. lintels, jambs and sills, should preserve the line of insulation, as illustrated in Diagram 3 of the Approved Document. An **alternative approach** would be to show by calculation that a satisfactory performance will be achieved, see Appendix D of the approved document.

Limiting infiltration of cold outside air via leakage paths should now be considered. These can have a significant affect on space heating demand, and any unintentional air paths should be reduced as far as practicable. The following measures would satisfy this requirement.

- The perimeter of dry-linings, including edges at openings and wall junctions, should be continuously sealed to the masonry wall with plaster or a similar continuous sealing method.
- Vapour control membranes in timber-frame constructions should be sealed.
- Windows, doors and rooflights should be fitted with draught seals and be sealed at frame perimeters.
- Loft hatches should incorporate draught seals.
- Service ducts and service pipes penetrating them should both be sealed, at floor, ceiling and entry positions as necessary.

The **control of space heating systems**, excluding individual heaters, is particularly important where appropriate provision should be made for:

- **Zone controls** to govern temperatures independently in separate areas: e.g. the living room(s) as one zone and the remainder of the house as the other zone. Room thermostats and/or thermostatic radiator valves, or any other suitable temperature-sensing devices, can be used for control. Ducted warm air systems and flap-controlled electric storage heaters need only thermostats.
- **Timing controls** to govern the duration of operation (not appropriate for systems with solid-fuel-fired boilers operated by natural draught).
- **Boiler control interlocks**, for gas and oil-fired systems, should switch off the boiler when there is no heat demand (unnecessary boiler cycling in systems with thermostatic radiator valves should be prevented).

To **control hot water storage systems**, excluding solid-fuel-fired systems, appropriate provision should be made for:

- a suitably sized **heat exchanger** in the storage vessel (cylinder), complying with BS 1566 or BS 3198 or equivalent;
- a **thermostat** to shut off the heat supply when the storage temperature is reached;

- a **timer** to shut off the heat supply for the periods when water heating is not required;
- a **thermostatically controlled valve** for a solid-fuel-fired system, if the cylinder does not give the slumber load.

As an **alternative approach** to the control of space and hot water storage systems reference may be made to BS 5449: 1990, BS 5864: 1989 or any other authoritative design specification recognized by the heating fuel supply company.

The **insulation of vessels, pipes and ducts** should be considered, based on the following guidance.

- Factory-applied insulation, e.g. 35 mm thick PU-foam (having zero ozone depletion potential), should be applied to hot water vessels to restrict heat losses to 1 W/l, based on a vessel with a 120 l capacity. For unvented hot water systems the safety fittings and pipework should also be insulated.
- Pipes and ducts should be insulated in unheated spaces, unless the heat loss contributes to the useful heat requirement of the room or space. Pipes, including those within 1 m of the cylinder, should utilize materials with a thermal conductivity not more than 0.045 W/mK and a thickness equal to the outside diameter of the pipe (maximum 40 mm) and 15 mm for the pipes within 1 m of the cylinder. Alternatively, for pipes and warm air ducts, refer to BS 5422: 1990. Note that additional insulation may be necessary to fully protect pipework from freezing.

Thermal insulation provisions to **conservatories** have often presented problems in the past, mainly with respect to what is or what is not a conservatory. Approved Document L now includes a definition, stating that, 'a conservatory has not less than three-quarters of the area of its roof and not less than one-half of the area of its external walls made of translucent material'. Translucent material would include glass and other transparent materials. Note that since a definition is not given in Schedule 2 – Exempt buildings and work, for conservatories less than 30 m^2, the one stated in this approved document could be used as a basis to establish whether a particular 'conservatory' is exempt or not. The guidance given for conservatories is for those attached to and built as part of a new dwelling, although it could also be applied to conservatories in excess of 30 m^2 which are not therefore exempt:

- A conservatory should be treated as an integral part of the dwelling if no separation exists between them.
- Where separation does exist between them, i.e. walls and floors to semi-exposed standards and windows and doors to the exposed

standard, then the conservatory may be heated, subject to the installation having its own temperature and on/off controls.

Section 2: Buildings other than dwellings

As with dwellings one of three methods can be used to demonstrate
compliance with Requirement L1:

- elemental method;
- calculation method;
- energy use method.

The **elemental method** requires the elements of construction,
namely walls, floors, roofs, windows, personnel doors, rooflights
and vehicle access doors, to meet the specified standard U-values.
A summary of the provisions, including allowances for windows,
personnel doors and rooflights, is illustrated in Figure 13.2. To
achieve the U-values stated in Diagram L2 reference can be made
to certified manufacturer's data or Appendix A of the approved

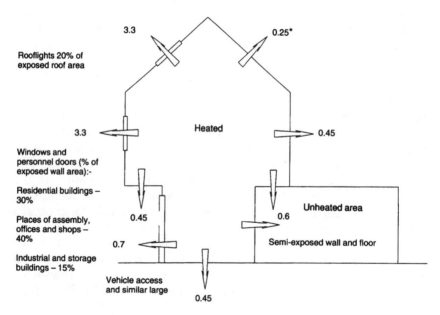

Figure 13.2 Standard U-values (W/m²K) and areas of openings for buildings
other than dwellings.
Notes: Places of assembly, offices and shops exclude display windows etc. A U-
value of 0.35 (residential buildings) or 0.45 (other buildings) applies to a flat roof
(or insulated sloping roof with no loft) as indicated by *.
A roof at a pitch of 70° or over may have the same U-value as a wall.

document. Appendices C and E contain examples of the procedures used for this method, including the calculation of U-values of ground floors.

The **basic allowance** for windows, personnel doors and rooflights, as stated in Figure 13.2, is seeking the provision of double glazing or secondary glazing to achieve a U-value of 3.3 W/m²K. Table 7 of the Approved Document gives indicative U-values for various window, door and rooflight constructions. Where enclosed, unheated and draughtproof lobbies are used to protect windows and personnel doors with single-glazed panels a U-value of 3.3 W/m²K can be assumed for these openings.

A **modification** to the basic allowances illustrated in Figure 13.2 can be made to allow larger areas of windows, personnel doors and rooflights. This is subject to improvements being made to the average U-value of the openings in question. Table 13.3 outlines the permitted range of variation.

The **calculation method** is similar to the target method used for dwellings, and can offer increased flexibility over the elemental method. The total rate of heat loss for the **proposed building** is calculated and then compared with a **notional building**, of the same size and shape, designed to the elemental method. That is:

Total rate of heat loss of the proposed building (W/K), area (m²) × U-value (W/m²K) for each element $<$ Total rate of heat loss of the notional building (W/K), area (m²) × U-value (W/m²K) for each element

Table 13.3 Permitted variation in the area of windows, doors and rooflights for buildings other than dwellings[a]

Average U-value (W/m²K)	Residential buildings (e.g. hotels) (% wall area)	Places of assembly, offices and shops (% wall area)	Industrial and storage buildings (% wall area)	Rooflights (all other buildings) (% roof area)
2.0	55	74	28	37
2.5	42	56	21	28
3.0	34	45	17	22
3.3	**30**	**40**	**15**	**20**
3.5	28	37	14	19
4.0	24	32	12	16
4.5	21	28	11	14
5.0	19	25	9	13

Note: For intermediate values reference should be made to Table 3 of the Approved Document.

Appendix H, Example 1, of the Approved Document provides example calculations.

The **energy use method** offers complete design freedom where the full range of valid energy conservation measures can be utilized, taking account of solar and internal heat gains. Similar to the calculation method, a comparison is made between the calculated annual energy use of the proposed building and that of a similar building designed to comply with the elemental method. This method would normally require the services of a **competent person** or an **approved person**. To demonstrate compliance for naturally ventilated buildings reference can be made to the method contained in the CIBSE publication *Building Energy Code* (1981), Part 2a (worksheets 1a to 1e)

When using the calculation and energy use methods the minimum U-values in residential buildings for exposed walls and floors should not be allowed to be worse than 0.7 W/m^{2}K and 0.45 W/m^{2}K for roofs. For non-residential buildings U-values of exposed walls, roofs and floors should not be worse than 0.7 W/m^{2}K.

As with dwellings, provisions need to be made to protect **thermal bridging locations** and to **limit the infiltration** of cold outside air via leakage paths. The guidance contained in Section 1 can be used for this purpose. Two minor additions are that hatches to unheated floor and roof voids should be draught sealed, and for larger more complex buildings reference can be made to BRE Report 265 *Minimizing air infiltration in office buildings* (1994).

The **control of space heating systems**, excluding commercial or industrial processes, should cover the following three areas to show compliance with the requirement:

- **Temperature controls** to govern temperatures independently for separate parts of the system. Thermostats and/or thermostatic radiator valves, or any other suitable temperature sensing devices, can be used for control. Hot water systems should incorporate an external temperature sensor and a weather compensator controller to regulate flowing water temperatures in the heating system.
- **Timing controls** to maintain the required temperature to each separate part of the system when the building is normally occupied. These controls could be: clock controls (system output less than 100 kW); optimizing controllers, which take account of the building cool-down/heat-up period (system output more than 100 kW); and any additional controls to prevent condensation or frost damage.
- **Boiler sequence controls** to detect variations in heating demand and start, stop or modulate boilers in combination for maximum efficiency (boilers serving loads more than 100 kW).

To **control hot water storage systems** the provisions outlined in Section 1 can be used.

The **alternative approach** for heating and storage system controls is to refer to BS 6880: 1988 *Code of practice for low temperature hot water heating systems of output greater than 45 kW* and CIBSE Applications Manual AM1: 1985 *Automatic controls and their implications for systems design*.

The **insulation of vessels, pipes and ducts**, excluding those used for commercial or industrial processes, should be based on the guidance given in Section 1, including the hot pipes connected to a hot water storage vessel complying with BS 853 or equivalent.

Requirement L1 (e) controls for the first time the efficiency and controllability of **lighting** in buildings, other than dwellings and where the floor area served exceeds 100 m². Lighting can consume significant levels of energy dependent on the efficacy of fittings and whether they are left on unnecessarily. The provisions made in the Approved Document seek to control these two aspects, although they do not apply to emergency/escape lighting or display lighting used to highlight exhibits or merchandise:

- At least 95% of the installed lighting capacity in circuit watts (power consumed by lamps and their control gear/power factor correction equipment) should be fittings with high-efficacy lamps, i.e. high-pressure sodium, metal halide, induction lighting, tubular fluorescent (25 mm diameter T8 lamps) or compact fluorescent (ratings above 11 W).
- The use of lamps with an average initial (100 hour) efficacy of not less than 50 lumens per circuit watt is given as an **alternative approach**.
- Lighting controls should encourage the best use of daylight but not endanger occupants using the building.
- Local switches (manual, remote or automatic) should be provided in accessible locations within working areas or on the boundary of working areas and circulation routes. As a guide the local switch should be within 8 m, or three times the fitting height if greater, from the furthest lighting fitting. Time switching and photoelectric switching may also be used as an **alternative approach** for buildings other than offices and storage buildings.
- For lighting controls the CIBSE publication *Code for interior lighting* could be used as an **alternative approach**, but designs should perform no worse than designs that follow the guidance on lighting generally in the Approved Document.

When undertaking a **material alteration**, to a dwelling or a building other than a dwelling, consideration should be given to the relevant requirements stated in Regulation 3, i.e. Part A, B1, B3, B4, B5 and Part M. In an effort to conserve energy the guidance contained in the Approved Document seeks to apply the provisions of Part L to a material alteration, dependent on the circumstances of the case. For example, the substantial replacement of a roof structure, involving Part A, may warrant the provision of satisfactory roof insulation to achieve the *U*-value for new dwellings. Whereas these increased thermal standards could be regarded as unreasonable if the roof structure were only modified. Three other situations are given:

- the provision of floor insulation where ground floors are substantially replaced in heated rooms;
- the provision of wall insulation where complete external walls are substantially replaced;
- the provision of controls and insulation where work is carried out on space/water heating systems.

A similar situation occurs with regard to a **material change of use**, although Requirement L1 *is* a relevant requirement and can therefore be applied in all cases, where the extent of provisions will depend on the circumstances of the case. In addition to those provisions stated under material alterations consideration should also be given to:

- the provision of additional insulation in accessible lofts to give a *U*-value of 0.35 W/m²K, where the existing is worse than 0.45 W/m²K;
- the provision of dry-lining where substantial areas of internal surfaces are to be renovated;
- the provision of replacement windows, draught-stripped and with a *U*-value of 3.3 W/m²K, unless town planning conservation restrictions apply.

In applying a reasonable degree of provisions for material alterations and material changes of use, careful and considered judgements will be necessary on the part of the designer and the building control authority.

APPENDICES

The principal contents of the appendices contained within the Approved Document have been touched upon within the body of the text; however, the appendix titles and the following specific comments have been highlighted for information.

Appendix A: Tables for the determination of the thicknesses of insulation required to achieve given *U*-values

Note that the values given in the tables have been derived from the proportional area method (see Appendix B).

Appendix B: The proportional area calculation method for determining *U*-values of structures containing repeated thermal bridges

The actual calculation method is given in the CIBSE Design Guide Section A3; this appendix gives worked examples for two external wall designs. The main factor to consider is the influence of the elements that present a cold bridge in the construction: e.g. the timber studs of the inner leaf of a timber frame construction, or the mortar joints within the inner block leaf of a traditional masonry cavity wall.

Appendix C: Calculation of *U*-values of ground floors

A ground floor of sufficient dimensions may give a *U*-value of 0.45 W/ m^{2}K or 0.35 W/m^{2}K without the provision of any insulation material. Figure 13.3 illustrates the limiting floor dimensions when insulation will be required. Appendix C also includes equations to establish the *U*-value of insulated and uninsulated floors. Further reference can also be made to BRE Information Papers 3/90, 7/93 and 14/94 (relating to basements).

Appendix D: Thermal bridging at the edges of openings

Procedures are given to establish whether there is an unacceptable risk of condensation at the edges of openings and/or whether the heat losses are significant. The minimum thermal resistance path is established, thermal resistances are calculated and then compared with satisfactory performance criteria to see whether action is necessary. An **alternative approach** is to refer to BRE Information Paper 12/94 *Assessing condensation risk and heat loss at thermal bridges around openings*.

Appendix E: The elemental method

A procedure is listed together with two worked examples.

Appendix F: The target *U*-value method

A procedure is listed together with two worked examples.

Appendix G: The SAP energy rating method for dwellings

The appendix, in its entirety, has been approved by the Secretary of State for the purposes of showing compliance with Regulation 16 for calculating SAP ratings for new dwellings. It is important that the

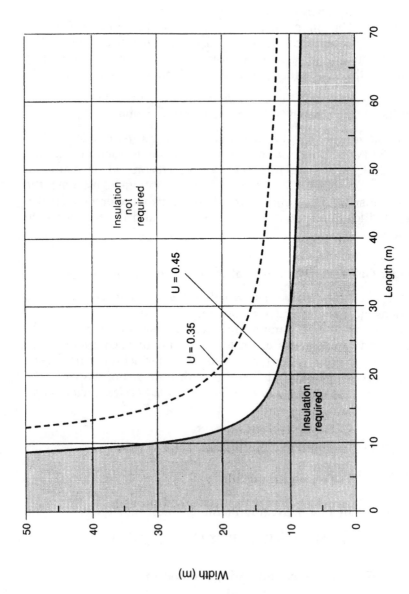

Figure 13.3 Floor dimensions for which insulation is required. Semi-detached, terraced or similar blocks of buildings can be taken separately or as one whole floor area.

calculation method, including the use of particular tabulated values, is used as set out in Appendix G. In due course the Secretary of State may approve modifications for the purposes of Regulation 16. A manual describing the Government's standard assessment procedure is included as part of the appendix and can also be obtained separately from the BRE. A SAP energy rating is based on the annual energy costs for space and water heating assuming a standard occupancy pattern and a standard heating pattern. The annual fuel cost per square metre of the floor area of the dwelling in question is expressed on a scale from 1 to 100: the higher the number the higher the energy efficiency of the dwelling, thus allowing potential home buyers to make a simple comparison between different new dwellings.

The SAP calculation is carried out on a Worksheet (Version 9.53) based on the BRE Domestic Energy Model (BREDEM). The numbered boxes are completed in sequence by referring to Tables 1–14 and the supporting guidance notes. A number of computer programs, approved by the BRE, are available to perform this time-consuming task. A computer spreadsheet can also be used to reproduce the content and result of the SAP worksheet.

Appendix H: Example calculations for buildings other than dwellings

Worked examples to calculate fabric insulation requirements and lighting efficiency are given.

CHAPTER 14

Approved Document M: Access and facilities for disabled people

In Great Britain over 6 million adults have a disability of one form or another. They may have difficulty in walking, need to use a wheelchair or suffer from a hearing or sight impairment. Requirements M2-M4 can be said to originate from the need to secure the welfare and convenience of persons in or about buildings, as prescribed in the Building Act 1984. With an ever-increasing awareness of the problems faced by disabled people, specific requirements were made, which have been expanded upon in the 2000 Building Regulations. The requirements now apply to new dwellings which have a significant impact on housing design.

Clarification is given of the relationship between the Disability Discrimination (Employment) Regulations 1996, and Part M. An employer will not be required to alter any physical characteristic included within a building which was adopted with a view to satisfying the requirements of Part M.

Similarly where compliance with Building Regulation requirement M2, in conjunction with Part K, is achieved. This is deemed to satisfy the relevant requirements of the Workplace (Health, Safety and Welfare) Regulations 1992, and the Health and Safety at Work, etc. Act 1974.

For the purposes of Approved Document M two specific definitions should first be studied.

REQUIREMENT M1: INTERPRETATION

In this Part, 'disabled people' means people who have:
(a) an impairment which limits their ability to walk or which requires them to use a wheelchair for mobility, or
(b) impaired hearing or sight.

REQUIREMENT M2: ACCESS AND USE

Reasonable provision shall be made for disabled people to gain access to and to use the building.

REQUIREMENT M3: SANITARY CONVENIENCES

(1) Reasonable provision shall be made in the entrance storey of a dwelling for sanitary conveniences, or where the entrance storey contains no habitable rooms, reasonable provision for sanitary conveniences shall be made in either the entrance storey or a principal storey.
(2) In this paragraph "entrance storey" means the storey which contains the principal entrance to the dwelling, and the "principal storey" means the storey nearest to the entrance storey which contains a habitable room, or if there are two such storeys equally near, either such storey.
(3) If sanitary conveniences are provided in any building which is not a dwelling, reasonable provision shall be made for disabled people.

REQUIREMENT M4: AUDIENCE OR SPECTATOR SEATING

If the building contains audience or spectator seating, reasonable provision shall be made to accommodate disabled people.
1. The requirements of this part do not apply to:
    (a) a material alteration;
    (b) an extension to a dwelling, or any other extension which does not include a ground storey.
    (c) any part of a building which is used solely to enable the building or any service or fitting in the building to be inspected, maintained or repaired.
2. Part M4 does not apply to dwellings.

Building – a building or a part of a building which may comprise individual premises; a shop, an office, a factory, a warehouse, a school or other educational establishment including student living accommodation in traditional halls of residence, an institution, or any premises to which the paying or non-paying public is admitted.

Principal entrance storey – means the storey level that contains the principal entrance(s) or the defined alternative accessible entrance.

The following meanings only apply to the sections on dwellings.

Common – serving more than one dwelling.

Habitable room – in relation to the principal storey, means a room used, or intended to be used, for dwelling purposes, including a kitchen but not a bathroom or a utility room.

Maisonette – a self-contained dwelling, but not a dwelling-house, which occupies more than one storey in a building.

Point of access – the point at which a person visiting a dwelling would normally alight from a vehicle which may be within or outside the plot, prior to approaching the dwelling.

Principal entrance – the entrance which a visitor not familiar with the dwelling would normally expect to approach or the common entrance to a block of flats.

Plot gradient – the gradient measured between the finished floor level and the point of access.

Steeply sloping plot – a plot gradient of more than 1 in 15.

Before moving onto the technical guidance contained in the Approved Document the first task must be to establish whether the requirements apply to the building work in question. The introduction of the Approved Document makes it clear that Requirements M2-M4 apply to:

- **New buildings**.
- **Buildings substantially demolished to leave only external walls**. In reconstruction, other than a dwelling, it may not be possible to provide for an accessible principal entrance; however, the remaining requirements should still be applied.
- **An extension incorporating a ground storey**. An extension independently approached and entered from the site boundary should be treated as a new building. Where access is via the existing building, improvements within it are not deemed necessary; any extension should be at least as accessible and usable as the extended building but should not adversely affect any existing disabled provisions.

Conversely, Requirements M2–M4 do not apply to:

- an extension that does not include a ground storey;
- a material alteration, unless the building will be rendered **less satisfactory** with respect to Part M, reference Regulations 3 and 4;
- a dwelling, or the common parts of a building used exclusively for two or more dwellings;
- a part of a building used only for inspection, maintenance or repair of the building itself or any service or fitting in the building;
- any external path or feature that does not form part of the external circulation or pedestrian access route to the building;
- buildings other than dwellings.

Once it has been established that the requirements will apply to the building, then reasonable provisions should be made to enable disabled people to:

Buildings other than dwellings

- reach the principal entrance(s), whether for customers, visitors or staff, from disabled parking areas and from the point of entrance to the site curtilage, avoiding elements of the building that could present a hazard for those with a sight impairment;
- gain access (approach or enter) into and within the building, to any storey level and to any facilities provided to accord with Part M;

- use sanitary accommodation, spectator seating and aids to communication, where needed for persons with sight or hearing impairments.

Dwellings (including any purpose built student accommodation, other than a traditional halls of residence providing mainly bedrooms and not equipped as self-catering accommodation)

- so that disabled people can reach the principal, or suitable alternative, entrance to the dwelling from the point of access;
- for access for disabled persons into and within the principal storey of the dwelling; and
- for sanitary accommodation at no higher storey than the principal storey.

Sections 1–10 tackle each of these items in a logical sequence.

Section 1: Means of access to and into the building other than dwellings

The provisions of this section recognize the differing needs of wheelchair users, ambulant disabled people and those with impaired sight, and are illustrated in Figure 14.1. The detailed design considerations are summarized for each element forming part of the route up to and into the building.

Where possible a **level approach** should be provided, which should be at least 1200 mm wide and not steeper than 1 in 20.

Pedestrian routes should incorporate dropped kerbs where necessary for wheelchair users and a tactile warning (blister-type paving) at

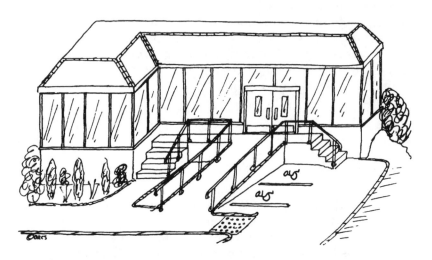

Figure 14.1 Principal provisions for approach up to building.

carriageway crossing points. For external steps the tactile warning (corduroy-type paving) should be provided at the top of the steps.

For an approach steeper than 1 in 20 a **ramped approach** should be adopted. This should incorporate suitable complementary steps, where practicable, based on the following guidance.

- It should provide a slip-resistant surface.
- It should be a minimum 1200 mm wide, 1000 mm minimum unobstructed width.
- It should not be steeper than 1 in 15 for individual flights up to 10 m long, or 1 in 12 for flights not exceeding 5 m.
- There should be landings to top and bottom of the ramp, minimum 1200 mm deep, and any intermediate landings 1500 mm deep, all clear of any door swing.
- Provide a minimum 100 mm kerb/upstand to open-sided flights and landings.
- Provide suitable continuous handrails to both sides of flights and landings where ramp is over 2 m long.

The provision of a **stepped approach** would normally be in addition to any ramped approach and should be based on the following guidance.

- Provide a tactile surface (corduroy-type paving) to the top landing, minimum 2000 mm x 800 mm, stopping 400 mm short of the first nosing and extending 150 mm either side of the stairs.
- Distinguish all nosings by the use of contrasting brightness.
- The steps should be a minimum 1000 mm unobstructed width.
- No rise between flights should exceed 1200 mm.
- Landings to top and bottom, and any intermediate landings, of the stepped approach should be a minimum 1200 mm deep, clear of any door swing.
- The maximum rise of the steps should be 150 mm (avoiding open risers), and the minimum going 280 mm.
- For two or more risers provide suitable continuous handrails to both sides of flights and landings.

Handrails for both ramped and stepped approaches should be grippable (45-50 mm diameter, 45 mm from any wall) and securely fixed at a height of 900 mm above the ramp surface or pitch line of steps, 1000 mm to landings. They should terminate in closed ends, extending 300 mm beyond the top and bottom of the ramp or step nosings, and not project into a route of travel.

People with sight impairments using access routes up to and around the outside of the building should be protected from potential hazards, such as opening windows and doors. Suitable guarding or handrails could be provided at door locations, while cobbles or a planting strip, and a slight change of level, could be used to identify the existence of windows.

The defined point(s) of access into the building for visitors, customers and staff (which may be specifically for their use) should all be suitable and accessible. Where sloping ground or space restrictions present problems at the principal entrance a suitable alternative entrance may be provided, subject to its being used for general use (i.e. not just a goods entrance) and that sufficient internal circulation exists in the building to allow access to the principal entrance storey. This principle also applies where access from disabled car parking cannot be achieved up to the principal entrance.

Principal entrance doors should allow wheelchair approach to enable the user to reach the door handle and give two-way visibility through the door to avoid collisions. The provisions should be as follows:

- minimum clear opening door leaf of 800 mm;
- minimum 300 mm unobstructed space on the leading edge side of the door, unless the door opens automatically;
- a visibility zone of glazing to cover a height of between 900 mm and 1500 mm above the floor level.

The provision of small revolving doors at the entrance should be avoided, but where installed a separate and suitable opening door should be provided. Some large revolving doors, however, do have sufficient internal space and the slow revolving speed necessary to allow use by wheelchair users.

Entrance lobbies should be made as generous as possible, minimum 1500 mm wide for single doors and between 1800 and 2400 mm deep depending on door swings. For 1800 mm double doorsets the depth of the lobby should be a minimum 2000 mm, for both sets of doors swinging in the same direction, or 2300 mm for double swing doors. Diagram 8 of the Approved Document refers.

Section 2: Means of access within buildings other than dwellings

Once access has been gained into the building consideration needs to be given to suitable internal circulation to the **principal entrance storey**, storey levels above and below this and the vertical circulation between them. Since the main guidance provisions take account of the space requirements for manoeuvring a wheelchair it should first be established which floor levels are to be accessible by wheelchair disabled people. The most straightforward method of vertical travel between floor levels is by way of a passenger lift, and for certain building purpose groups this may be required as a matter of course: for example, a residential care home or a purpose-built office building.

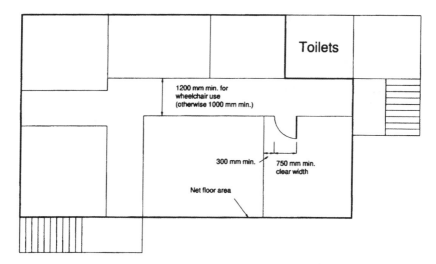

Figure 14.2. Calculation of net floor area of a storey. *Note:* Areas of a storey, served by the same entrance from the street (or mall), should be added together whether they are in different parts of the storey or are used for different purposes. Vertical circulation, sanitary accommodation and maintenance areas can be

However, having in mind the cost and floorspace implications of providing a suitable lift it is felt reasonable to only provide a **suitably designed passenger lift** to serve any storey above or below the principal entrance storey where the **net floor area** of that storey, as calculated from Figure 14.2, exceeds:

- 280 m², in the case of a two-storey building; or
- 200 m² for a building with more than two storeys.

Therefore, for a small three-storey residential care home or office, with net floor areas not exceeding 200 m² each, the provision of a lift would not be required from a building regulation point of view. If a lift was provided every opportunity should be made to make it suitable for use by disabled people, although this would not be controlled under Requirement M2. Where a suitable passenger lift is not installed a stairway designed for use by ambulant disabled people should be provided.

Having established which storeys require wheelchair access, consideration can now be given to **internal doors**, based on the following design guidance:

- minimum clear opening door leaf of 750 mm;
- minimum 300 mm unobstructed space on the leading edge side of

the door, unless the door opens automatically or where assistance is available;

- a visibility zone of glazing to cover a height of between 900 mm and 1500 mm above the floor level.

Corridors and passageways located in the principal entrance storey or in a storey served by a suitable passenger lift should be a minimum 1200 mm wide to allow wheelchairs to pass and manoeuvre. In all other situations a minimum width of 1000 mm should be adopted: this would also be acceptable where the building is only served by a stairway or for an extension that is approached via an existing building.

Internal lobbies should be made as generous as possible, minimum 1200 mm wide and between 1800 and 2400 mm deep, for single doors depending on door swings (see Diagram 10 of the Approved Document). Where double doorsets are to be utilized then the guidance for external lobbies should be applied.

For **vertical circulation** within the building a **suitable passenger lift** should be installed based on the net floor area served, as outlined above. Where a suitable lift is required (or is to be provided as an alternative to an ambulant stair) then the following design provisions should be followed.

- Allow a clear landing at each floor level of at least 1500 mm x 1500 mm.
- Provide a minimum 800 mm clear opening door width(s).
- Provide a minimum car dimension of 1100 mm x 1400 mm deep.
- Site landing controls between 900 mm and 1200 mm above floor level with tactile identification of floor level.
- Site car controls between 900 mm and 1200 mm above car floor level, a minimum 400 mm from front wall and with tactile identification of floor level selected if lift serves more than three floors.
- For a lift serving more than three storeys incorporate a visual/voice indication of floor level reached and a signalling system to allow a 5 second notification that a landing call has been answered and a 5 second dwell time (photo-eye or infrared door re-activating devices may override system subject to the door remaining fully open for 3 seconds).
- For full design details of a suitable passenger lift for disabled use reference should be made to BS 5655: Parts 1, 2, 5 and 7.

Where **internal ramps** are proposed, to gain access to different floor levels or storeys, then the guidance stated for an external ramped approach should be followed. The three remaining options for vertical circulation are illustrated in Figure 14.3.

For wheelchair access to **unique facilities** contained within a floor

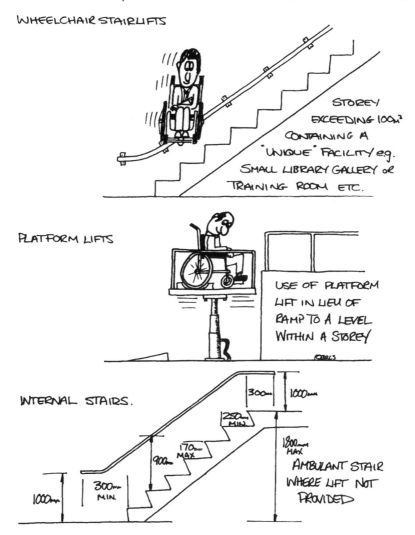

Figure 14.3. Remaining options for vertical circulation within the building.

area exceeding 100 m², such as those indicated in Figure 14.3, a **wheelchair stairlift**, to BS 5776: 1996, could be used where a passenger lift is impracticable to install.

Platform lifts, to BS 6440: 1999, generally allow wheelchair access between different levels within one storey, although a stair for ambulant disabled use should still be provided.

Where a suitable passenger lift is not provided, stairs should be designed to allow use by ambulant disabled people, based on the following guidance.

- Distinguish all nosings by use of contrasting brightness.
- Steps should be a minimum 1000 mm unobstructed width.
- No rise between flights should exceed 1800 mm (in exceptional circumstances the guidance in Approved Document K could be utilized).
- Landings to top and bottom, and any intermediate landings, of flights should be a minimum 1200 mm deep, clear of any door swing.
- The maximum rise of steps should be 170 mm (avoiding open risers), and the minimum going 250 mm.
- For two or more risers provide suitable continuous handrails to both sides of flights and landings, as detailed for a stepped approach.

Section 3: Use of buildings other than dwellings

Facilities incorporated into the building design should be both accessible and usable by disabled people. The guidance contained in the Approved Document does not address all possible facilities, dealing with only a few, and it may therefore be necessary to refer to other guidance sources to show compliance with the functional Requirement M2.

Common facilities, which include canteens and cloakrooms, doctors' and dentists' consulting rooms or other health facilities, should be located within a storey accessible by wheelchair users.

For **restaurants and bars** the guidance is that:

- suitable access should be made available to the full range of services;
- all bars and self-service counters and at least half of the seating provision should be accessible by wheelchair users; and
- different service areas, e.g. a combined self-service/waiter-service/take-away restaurant, should have wheelchair access to at least half of the areas in question.

Hotel and motel bedrooms should be suitable for wheelchair disabled people, who should also be able to visit other bedrooms, subject to suitable lift access being available to other storeys. The guidance is that:

- one guest bedroom in twenty (or part thereof) is suitable for wheelchair disabled use, with regard to size, layout, facilities and access;
- other guest bedrooms should also have a minimum 750 mm clear opening door leaf, although the 300 mm leading edge dimension need not be provided since assistance can be assumed.
- Diagram 13 of the Approved Document gives just one example of

an 'accessible' hotel bedroom and en-suite bathroom, which should have a minimum 1200 mm wide hallway, 1500 mm diameter wheel-chair turning circle in the bedroom and a 800 mm wide opening to the bathroom (if provided).

To enable the full use of swimming pools and other recreational build-ings, **changing facilities**, including shower compartments and dres-sing cubicles, should allow wheelchair-bound people to manoeuvre, transfer onto a seat and use taps, shower heads, mirrors and clothes hooks. For specific design guidance refer to Diagrams 14 and 15 of the Approved Document.

Aids to communication are particularly important to enable persons with impaired hearing to attend public performances or to take part fully in a discussion. These should be suitably planned and provided at ticket and booking offices (where a glass screen sepa-rates customer and vendor) and in large reception areas, in auditoria and meeting rooms over 100 m^2 in area. The two main systems in common use are:

- **loop induction** – where the amplified signal is directed through a loop around the relevant space and picked up by a listener's hear-ing aid, reference BS 6083; or
- **infrared** – which requires a personal receiver to demodulate and convert the signal in the invisible light.

Section 4: Sanitary conveniences in buildings other than dwellings

The availability of sanitary conveniences should in principle be the same for both disabled and able-bodied persons having regard to the overall building size and layout. The provisions made reflect the needs of disabled people in that they may require to reach a WC quickly, that WC compartments should offer ease of access at any time and take into account the differing circumstances of disabled persons at work or visiting the building.

Regardless of whether WC compartments, for wheelchair disabled use, are **unisex** (allowing assistance by a member of the opposite sex) or **integral** (contained within separate male and female accommoda-tion) they should satisfy the following design criteria:

- space for wheelchair manoeuvre;
- frontal, lateral, diagonal and backward transfer onto the WC;
- facilities to wash and dry hands within reach of WC; and
- sufficient space for a helper to assist in the transfer.

Sanitary conveniences provided for **visitors and customers** may be or include **unisex** compartments, suitable for wheelchair disabled use. For **staff**, sanitary conveniences may be **integral** or **unisex**. For a building with a lift this provision may be made on alternate floors, subject to a maximum horizontal travel distance of 40 m between a workstation and the WC, and that the general sanitary accommodation is located in an area that has unrestricted access. In a building without a suitable lift, WCs for wheelchair bound **staff** are required only to the principal entrance storey, unless the storey contains only the entrance or vertical circulation areas.

In addition to the provisions for visitors, customers and staff, a **hotel or motel** guest bedroom should contain a suitable **en-suite**, unless these are not generally provided, when a nearby **unisex** WC may be utilized.

A wheelchair WC compartment should be at least 1500 mm x 2000 mm deep and contain the necessary equipment and fittings as shown in Diagram 16 of the Approved Document. For buildings with more than one compartment, left-hand and right-hand transfer layouts should be adopted. It should also be noted that, subject to the compartment being able to comply with the design criteria stated above, reduced internal dimensions may be possible.

Turning to the needs of **ambulant disabled people**, at least one WC compartment, suitable for their use, should be included within each range of WCs for each storey where access is not possible for wheelchair users. This will be in addition to any provisions needed to the principal entrance storey. The WC compartment for ambulant disabled people should be at least 800 mm x 1500 mm deep, with handrails to both sides at a height of 700 mm (see Diagram 17 of the Approved Document).

Section 5: Audience or spectator seating in buildings other than dwellings

To give a clear view of a particular event in a theatre, cinema, concert hall, sports stadium, etc., and to allow a choice of seating location, six spaces or a 1/100th of the fixed seats provided should be available for wheelchair use. In large stadia a reduced percentage may be adopted.

The space allocated for a wheelchair should be a minimum 900 mm x 1400 mm deep, which may be created by the removal of seats, and dispersed among the remainder of the audience or spectator seating. The disposition and viewing locations in a theatre and a stadium area are shown in Diagrams 18 and 19 of the Approved Document, where specific consideration must be given to means of escape design (see Approved Document B).

Guidance on access for disabled people in sports stadia is included in the following:

- *Guide to Safety at Sports Grounds*
- *Designing for Spectators with Disabilities*
- *Access for disabled people*

Section 6: Means of access to and into the dwelling

Reasonable provision should be made within the boundary of the plot for disabled persons to approach and gain access into a dwelling from the point of alighting from a vehicle, which may be outside the plot. Note that the location and layout of dwellings/plots on a site will be a Town Planning matter.

This may be achieved by:

Approach to the dwelling

- A **level approach** with a gradient not exceeding 1:20 with a firm even surface, not loose, e.g. gravel, not less than 900 mm wide.
- A **ramped approach** which;
  - (a) has a surface which is firm and even.
  - (b) has flights of unobstructed width not less than 900 mm.
  - (c) has flights not longer than 10 m for gradients up to 1 in 15, or 5 m for gradients not steeper than 1 in 12.
  - (d) has top and bottom landings, and if necessary intermediate landings, with a length clear of any door or gate swings not less than 1.2 m.

Note – handrails and kerbs are not required, but see Approved document K where forming part of a building.

- A **stepped approach** where the point of access to the entrance has a gradient exceeding 1:15, which:
  - (a) has flights with unobstructed widths not less than 900 mm.
  - (b) has a rise of any flight between landing of not more than 1.8 m.
  - (c) has top and bottom and if necessary intermediate landings not less than 900 mm.
  - (d) has tread nosings as described for buildings other than dwellings, with the rise of each step being uniform and between 75 mm and 150 mm.
  - (e) has steps with goings not less than 280 mm, this is measured 270 mm from the "inside" for tapered treads, and
  - (f) has a suitable continuous grippable handrail (where three or more risers) on one side of the flight 850 – 1000 mm above the pitch line, and extend 300 mm beyond the top and bottom nosings.

- A **driveway** which provides an approach past any parked cars meeting the criteria of the above three methods of approach.

Access into the dwelling

Where the approach to the dwelling or block of flats is level or ramped an accessible threshold should be provided. An accessible threshold (showing compliance with Requirements C2 and C4) should also be provided to entrance level flats.

Where a stepped approach is unavoidable, an accessible threshold should be provided, and where a step into the dwelling is unavoidable the rise should not be greater than 150 mm.

An entrance door providing access for disabled people should have a minimum clear opening of 775 mm.

Section 7: Circulation within the entrance storey of the dwelling

Access should be facilitated within the entrance storey or the principal entrance storey into habitable rooms and a room containing a WC, which may be a bathroom on that level.

This is achieved through **corridors, passageways and internal doors within the entrance storey**, by way of doorways and corridor/passageways of minimum widths:

Doorway width	Corridor/passageway width
750 mm or wider	900 mm (when approached head-on)
750 mm	1200 mm (when approach not head-on)
775 mm	1050 mm (when approach not head-on)
800 mm	900 mm (when approach not head-on)

Local permanent obstructions, such as a radiator, are acceptable for a length not greater than 2 m providing the unobstructed width is not less than 750 mm for that length, and the obstruction is not placed opposite a door into a room so as to prevent a wheelchair user turning.

Where the plot is sloping a stepped change of level within the entrance storey may be unavoidable. **Vertical circulation within the entrance storey** may be achieved by a stair:

- with flights of clear width not less than 900 mm;
- a suitable continuous handrail on each side of the flight and any intermediate landings where there are three or more risers, and
- with rise and goings meeting the requirements of Part K for private stairs.

Section 8: Accessible switches and socket outlets in the dwelling

In order to assist those with limited reach wall mounted switches and socket outlets should be at suitable heights. That is outlets for lighting and other electrical appliances in habitable rooms should be between 450 mm and 1200 mm above finished floor level.

Section 9: Passenger lifts and common stairs in blocks of flats

Reasonable provision should be made to enable the disabled to visit occupants on any level. Ideally this should be by means of a lift, although this may not always be provided.

Where there is no passenger lift between storeys a stair suitable for use by the disabled should be provided, and in any event a stair within a common area should be suitable for the visually impaired.

Where a lift is provided it should be suitable for use by an unaccompanied wheelchair user, and those with sensory impairments.

Provisions for lifts

The provisions for lifts to dwellings are essentially the same as for other buildings as given is Section 2 with the exception of the lift car size. This should have a minimum load capacity of 400 kg and a width of at least 900 mm and length of at least 1250 mm (other dimensions may be used where test evidence, experience, etc. show that the lift will be suitable for unaccompanied wheelchair users).

Provisions for stairs

The provisions are similar to those of Section 2 except for:

- no minimum width is given (see Approved Document B),
- top and bottom landings to meet the requirements of Part K.

There is no requirement for intermediate landings as with other buildings.

Section 10: WC provision in the entrance storey of the dwelling

The entrance storey of a dwelling should be provided with a WC, which may be within a bathroom, and which is located so that it can be reached from a habitable room without the need to negotiate a stair.

If there are no habitable rooms in the entrance storey a WC may be provided in either the entrance storey or principal storey.

Any WC compartment should:

- have a door which opens outwards and have a width as given in Section 7 above, and
- provide a clear space for wheelchair users to access the WC, and hand basin positioned so as not to impede access.

Dimensional criteria is given for **frontal access** and **oblique access**. In each case there should be a clear area from the front edge of the WC pan of at least 750 mm deep, and:

- for frontal access, 450 mm minimum (500 mm preferred) width either side of the centreline of the pan, or
- for oblique access, 450 mm minimum (500 mm preferred) measured from the centreline of the pan on the door side, and 400 mm on the opposite side, and
- also for oblique access, the edge of the door opening closest to the WC cistern should be positioned 250 mm behind the front edge of the pan (measured perpendicular to the centreline of the WC pan. For clarification, please refer to Diagram 24 and 25 of the Approved Document.

In applying the guidance contained in Approved Document M, close cross-reference should be made to Approved Document B, concerning means of escape, and Approved Document K for the design and protection of stairs and ramps.

Finally, the Approved Document guidance does not give any **alternative approaches** to show compliance with Requirements M2–M4, except for sports stadia and educational establishments, where the guidance contained in Design Note 18 *Access for Disabled Persons to Educational Buildings* (1984), published by the Department for Education, may be used. It should be noted that this guidance does not confirm at which point a lift should be provided, but only states how that lift should be designed.

Some useful publications are listed for information:

- BS 5810: 1979 *Code of Practice for Access for the Disabled to Buildings.*
- PD 6523: *Information on access to and movement within and around buildings and on certain facilities for disabled people.*
- Department of the Enviornment, Transport and the Regions (1997) *Guidance on the Use of Tactile Paving Surfaces.*
- Selwyn Goldsmith (1997) *Designing for the Disabled: The New Paradigm*, RIBA.
- Centre for Accessible Environments: various publications, handbooks and design sheets available.
- *Accessible thresholds in new housing*, The Stationery Office 1999.

CHAPTER 15

Approved Document N: Glazing – safety in relation to impact, opening and cleaning

Requirements N1 and N2 seek to control the installation of glazing in **critical locations** in the building. A number of accidents have resulted in serious injuries and fatalities where people, both young and old, have fallen or walked into areas of glazing.

It is important to note that new glazing forming part of the erection, extension or material alteration of a building is defined as **building work**, and would thus need to comply with the two requirements. The installation of replacement glazing, on the other hand, is not controlled, although consumer protection legislation may apply. For an exempt extension, e.g. a conservatory not exceeding 30 m^2, a submission under the Building Regulations would not be necessary, although the glazing used should comply with Requirement N1, and N2 if applicable.

Requirement N3 and N4 were introduced in January 1998 and deal with the safe cleaning, opening and closing of windows, etc. The Workplace (Health, Safety and Welfare) Regulations 1992 – main requirements, relating to building design, are now covered by the Building Regulations. Compliance with Requirements N3 and N4, in accordance with Section 23(3) of the Health and Safety at Work, etc. Act 1974, would prevent the service of an improvement notice under the Workplace Regulations. For further information the Approved Document also makes reference to *Approved Code of practice and Guidance;*The Health and Safety Commission, L24 (HMSO 1992).

As with Part K, for mixed use developments (including dwellings) the requirements for non-domestic use can be applied to the shared parts of the building.

REQUIREMENT N1: PROTECTION AGAINST IMPACT

Glazing, with which people are likely to come into contact whilst moving in or about the building, shall:

(a) if broken on impact, break in a way which is unlikely to cause injury; or

(b) resist impact without breaking; or

(c) be shielded or protected from impact.

REQUIREMENT N2: MANIFESTATION OF GLAZING

Transparent glazing, with which people are likely to come into contact while moving in or about the building, shall incorporate features which make it apparent.

Requirement N2 does not apply to dwellings.

The guidance contained in the Approved Document is split into four sections, dealing with Requirements N1, N2, N3 and N4 in turn. Reference to Approved Documents B and K may also be necessary if the glazing performs a fire-resisting function or if it offers protection or containment to stairs or ramps.

Section 1: Protection against impact

To limit the risk of people in or about the building sustaining cutting or piercing injuries, certain **critical locations** are identified. These are illustrated by the shaded areas in Figure 15.1, and recognize the potential areas of impact at door positions and at low level, where children are particularly at risk.

To show compliance with Requirement N1, one of four options for glazing in critical locations may be chosen.

Safe breakage of a glazing material is defined in BS 6206: 1981 *Specification for impact performance requirements for flat safety glass and safety plastics for use in buildings.* For panes in doors, and door side panels more than 900 mm wide, a Class B material should be used as a minimum. For other critical locations a Class C material may be used, noting that Class A to BS 6206 is the highest. This classification, together with the material code, BS number and product name or trademark should be permanently marked on the material.

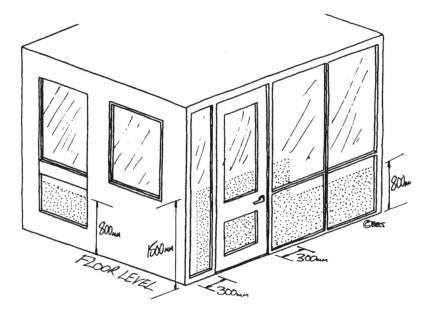

Figure 15.1.   Critical locations in internal and external walls.

Robustness is an inherent quality of a range of glazing materials, i.e. polycarbonate sheet or glass blocks. Annealed glass gains strength with thickness and can be used in large areas, e.g. shopfronts, although it should not be used in doors. As a guide, 10 mm annealed glass can be used for a maximum pane size of 2.25 m square, whereas no limits apply to a thickness of 15 mm.

Glazing in small panes relates to the use of a number of small isolated panes and includes traditional leaded lights or copper lights. The pane should not have a width exceeding 250 mm or have an area greater than 0.5 m^2. Annealed glass should be a minimum of 6 mm thick (4 mm in leaded or copper lights).

Permanent screen protection calls for the provision of a robust screen, which should prevent a 75 mm sphere from coming into contact with the glazing, where the glazing itself would not need to satisfy Requirement N1. If the screen also serves as guarding (see Approved Document K), then it should be unclimbable.

Section 2: Manifestation of glazing

Critical locations with respect to this requirement include large uninterrupted areas of transparent glazing, e.g. internal office screens or external walls to shops, factories, public buildings, etc. These elements, especially if located on floors that are at the same level,

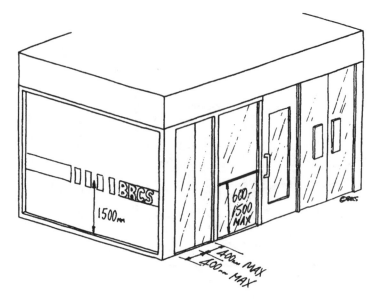

Figure 15.2. Methods to indicate large areas of transparent glazing.

may give the impression that access is possible from one part of the building to another without interruption. To avoid impact in such circumstances glazing should have permanent **manifestation** (in the form of lines, patterns or logos) or some other means to indicate the existence of these large areas of glazing. Figure 15.2 illustrates where manifestation should be provided and where it would not be required owing to the design of doors or the size of glazing panels.

REQUIREMENT N3: SAFE OPENING AND CLOSING OF WINDOWS, ETC.

Windows, skylights and ventilators which can be opened by people in or about the building shall be so constructed or equipped that they may be opened, closed or adjusted safely.
Requirement N3 does not apply to dwellings.

Section 3: Safe opening and closing of windows, etc

To comply with this requirement, some options are described. It should be noted that no reference is made to the safe use of windows, etc, by disabled people, this is not an issue controlled in Approved Document M.

- Height of controls should not be more than 1.9 m where unobstructed.
- If obstructed, the height of the controls should be lowered; for example, 1.7 m where an obstruction is 900 mm high and 600 mm deep.
- Where controls cannot be suitably positioned, then consideration will need to be given to remote operation (manual or electrical). This may be appropriate for disabled people.
- Where there is a danger of falling through a window (above ground level) then suitable opening limiters or appropriate guarding installed (cross reference Approved Document K).

REQUIREMENT N4: SAFE ACCESS FOR CLEANING WINDOWS, ETC.

Provision shall be made for any windows, skylights, or any transparent or translucent walls, ceilings or roofs to be safely accessible for cleaning.

Requirement N4 does not apply to:
(a) dwellings, or
(b) any transparent or translucent elements whose surfaces are not intended to be cleaned.

Section 4: Safe access for cleaning windows, etc.

A series of options to allow safe cleaning of both sides of the glazing where someone could fall more than 2 m are described.

- Use a suitably designed window that allows the outside surface to be cleaned from the inside. Diagram 8 of the Approved Document gives safe reach dimensions and further reference can be made to BS 8213: Part 1 *Windows, doors and rooflights*.
- Use of a portable ladder, not more than 9 m long, placed on an adequate area of firm ground, 75° pitch and provided with suitable tying/fixing point if over 6 m long.
- Use of minimum 400 mm wide walkways with 1100 mm high guarding or anchorage for sliding safety harness.
- Use of access equipment (with safety harnesses), for example; suspended cradle, travelling ladders or 'cherry picker'.

- Use of suitable anchorage points for safety harnesses/abseiling hooks.
- As a last resort, allow sufficient space for the erection of a temporary tower scaffold.

Further information

ADDRESSES

Architects and Surveyors Institute (ASI)
St Mary House
15 St Mary Street
Chippenham SN15 3WD

Tel: 01249 444505   www.asi.org.uk

Association of Builders' Hardware Manufacturers (ABHM)
42 Heath Street
Tamworth B79 7JH

Tel: 01827 52337   www.abhm.org.uk

Association of Building Engineers (ABE) [and The Association of
Corporate Approved Inspectors]
Lutyens House
Billing Brook Road
Weston Favell NN3 8NW

Tel: 01604 404121   www.abe.org.uk [www.acai.org.uk]

Association of Specialist Fire Protection (ASFP)
Association House
235 Ash Road
Aldershot GU12 4DD

Tel: 01252 321322   www.asfp.org.uk

British Board of Agrément (BBA)
PO Box 195
Bucknalls Lane
Garston
Watford WD2 7NG

Tel: 01923 665300   www.bbacerts.co.uk

British Gas
17 London Road
Staines TW18 4AE

Tel: 01784 645 000   www.gas.co.uk

British Standards Institute (BSI)
BSI Customer Services, Publications
389 Chiswick High Road
London W4 4AL

Tel: 0208 996 7000   www.bsi-global.com

Building Research Energy Conservation Unit (BRECSU)
Enquires Bureau
Building Research Establishment
Gaston
Watford WD2 7JR

Tel: 01923 664258   www.energy-efficiency.gov.uk

Building Research Establishment (BRE) [and the Fire Research Station
(FRS) and Loss Prevention Council (LPC)]
Garston
Watford WD2 7JR

Tel: 01923 664000   www.bre.co.uk

Building Services Research and Information Association (BSRIA)
Old Bracknell Lane West
Bracknell RG12 7AH

Tel: 01344 426511   www.bsria.co.uk

Centre for Accessible Environments (CAE)
Nutmeg House
60 Gainsford Street
London SE1 2NY

Tel: 0207 357 8182   www.cae.org.uk

Chartered Institute of Building (CIOB)
Englemere
Kings Ride
Ascot SL5 8BJ

Tel: 01344 630700   www.ciob.org.uk

Chartered Institution of Building Services Engineers (CIBSE)
Delta House
222 Balham High Road
London SW12 9BS

Tel: 0208 675 5211   www.cibse.org

Construction Industry Council (CIC)
26 Store Street
London WC1E 7BT

Tel: 0207 637 8692   www.cic.org.uk

Construction Industry Research & Information Association (CIRIA)
6 Storey's Gate
London SW1P 3AU

Tel: 0207 222 8891   www.ciria.org.uk

Department of Education and Employment (DfEE)
Sanctuary Buildings
Great Smith Street
London SW1P 3BT

Tel: 0870 000 2288   www.dfee.gov.uk

Department of the Environment, Transport and The Regions (DETR)
[Department for Transport, Local Government and the
    Regions (DTLR) from July 2001]
3rd Floor
Elland House
Bressenden Place
London SW1E 5DU

Tel: 0207 890 3000   www.detr.gov.uk [dtlr.gov.uk]

Environment Agency
Rio House
Waterside Drive
Aztec West
Almondsbury
Bristol BS32 4UD

Tel: 01454 624 400   www.environment-agency.gov.uk

Fire Protection Association (FPA)
Bastille Court
2 Paris Gardens
London SE1 8ND

Tel: 0207 902 5300 www.thefpa.co.uk

Health and Safety Executive (HSE)
Information Centre
Broad Lane
Sheffield S3 7HQ
(Postal enquires)

Telephone enquires: Tel: 08701 545500 www.hse.gov.uk

The Stationery Office
Publications Centre
PO Box 276
London SW8 5DT

Tel: 0207 873 9090 (orders); 0207 873 0011 (enquiries)
www.thestationeryoffice.com

Institute of Building Control (IBC) [see RICS]

Institute of Clerks of Works of Great Britain Incorporated (ICW)
41 The Mall
Ealing
London W5 3JT

Tel: 0208 579 2917 www.icwgb.com

Institute of Wastes Management (IWM)
9 Saxon Court
St Peter's Gardens
Northampton NN1 1SX

Tel: 01604 620426 www.iwm.co.uk

Institution of Civil Engineers (ICE)
1 Great George Street
London SW1P 3AA

Tel: 0207 222 7722 www.ice.org.uk

Institution of Fire Engineers (IFE)
148 New Walk
Leicester LE1 7QB

Tel: 0116 2553654   www.ife.org.uk

Institution of Structural Engineers (ISE)
11 Upper Belgrave Street
London SW1X 8BH

Tel: 0207 235 4535   www.istructe.org.uk

National House Building Council (NHBC)
Buildmark House
Chiltern Avenue
Amersham HP6 5AP

Tel: 01494 434477   www.nhbc.co.uk

Oil Firing Technical Association for the Petroleum Industry (OFTEC)
Century House
100 High Street
Banstead SM7 2NN

Tel: 01737 373311   www.oftec.org

Royal Institute of British Architects (RIBA)
66 Portland Place
London W1B 1AD

Tel: 0207 580 5533   www.architecture.com

Royal Institution of Chartered Surveyors (RICS) [and RICS Building
Control Forum formerly IBC]
12 Great George Street
London SW1P 3AD

Tel: 0207 222 7000   www.ricson line.org

Steel Construction Institute (SCI)
Silwood Park
Ascot SL5 7QN

Tel: 01344 623345   www.steel-sci.org

Timber Research and Development Association
    Stocking Lane
    Hughenden Valley
    High Wycombe HP14 4ND

    Tel: 01494 563091   www.trada.co.uk

United Kindgom Accreditation Service (UKAS)
    21–47 High Street
    Feltham
    Middlesex TW3 4UN

    Tel: 0208 917 8400   www.ukas.com

Zurich Municipal
    Southwood Crescent
    Farnborough GU14 ONJ

    Tel: 01252 522000 (Marketing Division)
    Tel: 0541 545500 (books)

Index